OBSESSION

A Study of Male Body Obsession in the USA

OBSESSION

A Study of Male Body Obsession in the USA

David Metcalfe BSc(Hons)

Warwick Medical School
University of Warwick, Coventry, UK

ISBN 978-1-84799-860-6

Contents

Acknowledgements

This is the second occasion on which I have cause to thank the committee and trustees of the Lord Rootes Memorial Fund. It is a rare event that an undergraduate is offered the means to direct and pursue their own research overseas. Nevertheless, the Lord Rootes Memorial Fund has now offered me this opportunity on at least two separate occasions and, as such, has directly informed the present work through its patronage. I would also like to thank the Lord Rootes Memorial Fund for honouring my project with an additional prize which first raised the possibility of publication. A further thank you is due to Prof. Yvonne Carter, Dean of the Warwick Medical School, whose Dean's Fund has contributed greatly towards the costs associated with publishing this study. Furthermore, I could not have produced this book without the help of Mina Aletrari who has proof-read almost everything that I have ever written.

I would also like to thank Prof. Harrison Pope of Harvard Medical School for providing comprehensive access to all of his papers on male body image.

Finally, a huge thank you to the many men in Southern California and beyond who suffered my questions both in the gym and on the beach with varying degrees of good humour.

Forward

The present monograph has been written in two parts – each describing an entirely separate phenomenon; one of which must nevertheless inevitably follow the other. The first part describes body image concerns as they are experienced by ordinary boys and men. The focus here is particularly on concerns surrounding muscularity. I have not restricted myself to merely describing the extent of these concerns and attempt to attribute causation where possible.

The second part, on the other hand, describes body image concerns as they are experienced only by a minority of men. These men are significant; however, as their preoccupation with appearance concerns have led them to psychosocial damage or to illegal performance enhancing drugs. Alternatively, their body image concerns have developed into, or have manifested themselves as, full blown psychiatric disorders. They suffer from potentially lethal pathologies such as anorexia, bulimia and body dysmorphic disorder (BDD). As the focus of this report is on muscularity, special consideration is shown to one particular branch of BDD – that of muscle dysmorphia.

My concluding section attempts to justify this organization by considering how body image concerns in ordinary men might be related to the obsessive behaviours and distorted perceptions of self experienced by the psychiatrically ill.

Introduction

Millions of men in the United States have a secret obsession: they are preoccupied with the appearance of their bodies. They are worried that their body is inadequate; their skin scarred, genitals too small or their torso insufficiently muscular [Pope *et al*, 2002]. Indeed, American men only just lag behind women in worrying about their looks with 82 percent actively trying to improve their appearance [Clausen, 2005].

The number of men dissatisfied with their appearance has risen by 300 percent in the last 25 years [Phillips & Castle, 2001]. The results of this shift have been profound. American men now spend over $4 billion a year on gym memberships and almost the same again on cosmetic toiletries such as moisturisers and tooth whiteners. These figures do not include the billions of dollars spent on elective surgical procedures. In 1996 alone, men in America received almost 700,000 cosmetic surgical operations ranging from muscle implants to penis augmentations to face lifts [Pope *et al*, 2002].

Appearance concerns are most likely to affect young men. Almost seven percent of American boys in secondary school admit to having used illegal anabolic steroids in order to increase muscularity [Phillips & Castle, 2001]. These illegal drugs have been implicated in mood disorders, liver damage and a heightened risk of prostate cancer [Embleton & Thorne, 1998]. Indeed, seventeen percent of males surveyed by an American magazine said they would sacrifice three years of their lives to reach their target body shape [Corson & Andersen, 2002].

Some have taken their appearance concerns to even greater extremes. Body dysmorphic disorder (BDD) is a pathology in which the patient becomes preoccupied with a non-existent or minimal cosmetic defect [American Psychiatric Association, 2000]. Sufferers of BDD are likely to become reclusive, self-harm or attempt suicide [Phillips, 1996]. It is thought that some four percent of American college students suffer from BDD, which affects as many men as it does women [Bohne *et al*, 2002]. One common form of BDD is muscle dysmorphia in which the patient, typically a man, unrealistically perceives himself to be inadequately muscular [Olivardia *et al*, 2000].

In Greek mythology, Adonis was a young man so handsome that the goddess Aphrodite herself fell in love with him [Burkert, 1985]. For this reason, Pope *et al* have dubbed the striving for body perfection among contemporary men as the 'Adonis Complex' [Pope *et al*, 2002]. This disorder is considered, by many clinicians, to differ from body dysmorphic disorder only in terms of degree. Just as feminists did for women in the 1970s, Pope *et al* explain rising male insecurity as a result of unachievable body types being portrayed by the media. The resulting Adonis Complex, they believe, manifests itself in men who strive to improve their appearance while simultaneously risking their health. This study concerns itself with a qualitative and quantitative exploration of the Adonis Complex and its impact on contemporary men in the United States.

Methodology

THE LOCALITY

The fieldwork for this study took place in and around the district of Venice in Los Angeles, USA. The reasons for this choice of locality were both historical and practical in nature. Venice is, for example, the site of the infamous Muscle Beach; an outdoor gymnasium said to have initiated the bodybuilding phenomenon which continues to sweep across the Western world today. From Muscle Beach came such celebrities as Arnold Schwarzenegger who was recently elected Governor of California.

A view overlooking the Venice promenade

The district of Venice also gave rise to the two largest fitness centre chains in the United States, namely Gold's Gym and World Gym. Indeed, Gold's Gym was featured in the movie *Pumping Iron* as early as 1977. For these historical reasons, the areas around Venice and nearby

Santa Monica have achieved something akin to cult status among contemporary bodybuilders. One consequence of this status is that Venice continues to attract an eclectic and cosmopolitan mix of both amateur and competitive bodybuilders. Many of these men were invited to participate in this study of the issues surrounding male body obsession in the United States.

The original Gold's Gym (right) and its contemporary replacement (left)

UNSTRUCTURED INTERVIEWS

Much of the data presented in this work was obtained by means of unstructured interviews. Many previous researchers have recruited study participants on their way into gyms; an approach considered to be inadequate for this project as only a minority of men entering gyms can be considered serious weight lifters. Instead, participants were recruited primarily on Venice Beach and selected for their high degree of muscularity. To qualify as a participant, individuals had to work out, or engage in a sport aimed primarily at improving muscularity, at least twice a week. In terms of the interview itself, this typically took place in the form of a general conversation in which opinions were solicited based on statements made my previous interviewees. Men differed widely in their initial response to my approach. Their responses varied from suspicion to lack of interest to great enthusiasm for talking about their pastime. This reaction usually informed the duration of the interview with only a few questions asked of uninterested men but long dialogues held with their more talkative counterparts. In some cases the 'interview' was held with friends or training partners present. While some responses might have been restricted in such a public forum, these interviews were more interactive and sometimes facilitated a more in-depth analysis of bodybuilding culture.

Lines of questioning typically sought to explore experience of weight lifting, use of legal body image drugs and other potential risk factors such as use of the internet and competition bodybuilding. Although use of illegal drugs was not asked about directly, this line of questioning was pursued when first raised by the interviewee. Brief notes were made from memory following each interview in order to inform the present report. Interview responses are presented here in an almost wholly qualitative manner to illustrate key themes raised and to further explore questionnaire findings.

QUESTIONNAIRES

Interviewees were additionally given a website address allowing access to an online anonymous questionnaire. This address was also marketed to men who were uninterested or who did not have the time to complete an interview. Similarly, forum participants were asked to consider completing the questionnaire. The online questionnaire facilitated a quantitative analysis of the information collected during interviews. In general this was considered to be a superior mode of data collection as respondents were more likely to answer sensitive questions honestly than in an interview situation with a stranger. Interviewees were often, for example, eager to downplay the amount of effort invested in weight lifting and to deny using nutritional supplements.

Two anonymous online questionnaires were used in total. The first applied to all of the men interviewed and requested general information about their patterns of weight training behaviour. 228 men completed this questionnaire, the results of which may be found in Appendix I. The second anonymous online questionnaire solicited information about illicit anabolic-androgenic steroid (AAS) use which could not be adequately covered in the majority of interviews. In order to maximize the number of respondents, this questionnaire was also heavily marketed to men using internet forums to discuss AAS use. 91 men in total completed this second questionnaire, the results of which are presented in Appendix II.

ONLINE FOCUS GROUPS

Online focus groups were used to explore the experiences of amateur weight lifters and patients diagnosed with body dysmorphic disorder (BDD). These often took place at online internet forums. Discussions with the former group were used as a preliminary measure to inform the structured interviews described above. Further focus groups were used after the fieldwork in America in order to clarify key points and to confirm my understanding of important issues. A focus group was, for example, employed to study the revelation that a large number of bodybuilding men would sacrifice years of life for their ideal body. These discussions often strayed from the original topic, however, and occasionally produced interesting details of their own. Furthermore, many of the images and details of illegal drug products were provided courtesy of steroid users recruited during these focus groups. Permission was obtained for reproducing these images although, for obvious reasons, full credit cannot be given in this publication.

PERSONAL CORRESPONDENCE

Patients diagnosed with body dysmorphic disorder (BDD) were also recruited from an online internet forum. Correspondence continued by email to aid an understanding of the nature of BDD from the perspective of those who suffer from it. These data were then used to explore the relationship between ordinary body enhancement, BDD and additional psychiatric disorders.

Ethical Considerations

CONFIDENTIALITY

In the process of conducting this investigation, particular attention was paid to safeguarding the confidential nature of interview responses. Although interviews were held in what were essentially public places, care was taken not to pursue difficult subjects within earshot of other people. In particular, participants were not asked directly about their use of illegal drugs; although a number of men did raise this issue of their own accord. Although there is an ethical obligation to breach confidentiality under certain circumstances, i.e. if a disclosure revealed the participant to be in imminent danger, use of anabolic steroids would not constitute 'imminent danger' in this instance. In general, however, information about AAS use was confined to the online questionnaires which guaranteed complete anonymity to respondents. Notes taken to record the outcome of interviews on Venice Beach were strictly anonymised so that confidentiality could be guaranteed even in the in the event of loss or theft.

During online focus groups and email correspondence with study participants, questions did not solicit personal details which might lead to identification of the individual concerned.

SENSITIVITY

Sensitivity was particularly important during focus groups and correspondence with BDD sufferers. Many of these respondents suffered from psychiatric disorders such as chronic depression and schizophrenia in addition to BDD. Many further had histories of self harm and/or attempted suicide. In light of this, particular care was taken not to probe directly into these sensitive issues. Questions were instead asked about BDD in general so that respondents were free to draw upon their own experiences, the experiences of others or simply upon what they had read around the subject.

Respondents were, without exception, keen to discuss their condition with a non-sufferer. This willingness appeared to stem from the perception among sufferers that BDD is an under-recognized and poorly understood disorder in the wider community.

P A R T 1

The Rise of Male Body Dissatisfaction

Part I

1.1

Body Image Concerns in Men

Almost all men harbour concerns about their appearance [Phillips, 1996]. In one survey of 30,000 *Psychology Today* readers, only eighteen percent of men indicated that they would not choose to improve their appearance. Furthermore, a full 33 percent of male respondents indicated that they would consider cosmetic surgery [Cash *et al*, 1986]. Mild appearance concerns are perhaps understandable as all available evidence suggests that physical attractiveness results in more successful interpersonal relationships and greater career success [(Cash *et al*, 1977) & (Clark & Mills, 1979)]. Indeed, one survey has suggested that unattractive men earn on average fifteen percent less than those who perceive themselves to be more attractive [Harper, 2000]. In line with these benefits, however, longitudinal analysis of male responses to body dissatisfaction surveys suggests that appearance concerns are increasing at an alarming rate [(Berscheid *et al*, 1973) & (Cash *et al*, 1986)]. Men who are very concerned about aspects of their appearance are most likely to worry about their height, skin, facial appearance, hair thinning, genitals and body muscularity [Raevuori *et al*, 2006].

1.1.1

HEIGHT

Although over ten percent of men are concerned about their height, only four percent of body dysmorphic men exhibit preoccupation with this feature of their body [Phillips & Diaz, 1997]. Unlike some other components of body image, men are often able to judge their own height accurately. Men shorter than 174cm have, for example, been found to be considerably more concerned about their height (OR 3.9 CI 2.9-5.2) than taller men [Raevuori *et al*, 2006]. Height concerns are, then, more likely to afflict men of ethnic groups associated with shorter stature. As many as 7.3 percent of Korean boys have, for example, been found to use herbal medicines, growth-promoting supplements or human growth hormone in order to increase their height [Park *et al*, 2003]. Height concerns are likely to have profound implications on the confidence and self-worth of men whom they afflict. One 170cm eighteen year-old Hispanic man in

LA, for example, asked whether taking anabolic-androgenic steroids (AAS) might help increase his height. He felt strongly that both men and women took him less seriously than other men because of his stature. Indeed, the height concerns of shorter men may in some cases be warranted. One survey, for example, has shown that short men earned ten percent less and are seven percent less likely to be married by age 33 than their otherwise identical but taller counterparts [Harper, 2000].

1.1.2
FACIAL APPEARANCE

As a short, uncontrolled experiment towards the end of this project, the most forthcoming interviewees were asked to rate their facial attractiveness by marking a score out of ten on a sheet of paper. A total of 57 weight lifters agreed to complete this task. This investigation is of limited use largely because it cannot be compared to a control group of men which do not regularly weight train. The average self-rating was, however, only 6.02 which indicates that many weight training men feel insecure about their facial appearance. Furthermore, many appeared to have found this numerical scoring system to be inadequate and had made additional notes on the form. One man for example wrote that he had *"never had trouble with the ladies... but it can't have anything to do with my face... I have a huge nose"*. Others wrote *"no idea how I get girls!"* and, perhaps more alarmingly, *"I have nice looking hair...so I tell girls just to look at that and try and ignore the face"*. Others noted conditions or provisos which must be met in order for them to justify the rating they had assigned themselves. These included comments such as *"after half an hour of grooming"*, *"on a good day"* and *"from the front.. my side profile looks horrible"*.

Dissatisfaction with facial appearance can have significant implications for confidence and self-perception. The *New York Times*, for example, has described the case of a young man who left his job and stayed at home for fear that people were staring at his *"pointed ears"* and *"large nostrils"* [Goleman, 1991]. Table 1 further shows that many body dysmorphic men are pathologically obsessed with various aspects of facial appearance.

Table 1: Prevalence of facial complaints in men with body dysmorphic disorder [Phillips & Diaz, 1997]

Facial feature	Proportion
Nose	38%
Eyes	18%
Lips	11%
Ears	9%
Face shape or size	8%
Chin	7%
Teeth	7%
Head shape or size	6%
Cheeks	5%
General ugliness	4%
Forehead	3%
Eyebrows	1%

1.1.3
SKIN

After nose complaints, the notes made by men rating their facial appearance most commonly mentioned skin condition. In particular men wrote about *"oily skin"*, *"pimples"* and *"acne scars"*. Although research indicates that the lives of younger men are less affected by acne than those of their older counterparts, social functioning may be significantly impaired by prominently visible skin complaints [Lasek & Chen, 1998]. Indeed, one investigation has shown that young acne sufferers exhibit a heightened prevalence of suicidal thoughts compared to general medical patients [Gupta & Gupta, 1998].

1.1.4
HAIR THINNING

57 percent of body dysmorphic men report an obsessive preoccupation with their hair. This typically centres around androgenetic alopecia, also known as hereditary hair loss. Around 47 percent of European men

between eighteen and 45 report having experienced hair loss which significantly alters self-image and psychosocial functioning [Alfonso *et al*, 2005]. The psychosocial trauma associated with hair loss has been shown to be particularly acute when it is first noticed or diagnosed [Passchier *et al*, 2006]. Men associate hair loss with personal unattractiveness (43%), fear of becoming bald (42%), becoming older (37%), impact of social life (22%) and depression (21%) [Alfonso *et al*, 2005]. At their most extreme, balding men may suffer a complete crisis in their self confidence. As one patient has described his hair loss, *"bald guys are total rejects... my life will be over when I lose all my hair"* [Phillips, 1996]. In order to slow or hide hair loss, many men take prescription medications such as propecia, use camouflage, hairpieces or comb-over techniques, or seek surgical hair transplantation.

1.1.5
GENITALS

Compared to other primates, including the 400lbs gorilla, human male genitalia are remarkably large [Small, 1996]. Despite the fact that penis size is similar for most adult males, men consistently underestimate the size of their own genitals. Concerns about penis size are most acutely felt during adolescence but often also persist into young adulthood [Lee & Reiter, 2002]. These concerns are usually associated with the perception that penis size is a mark of masculinity or may impact on sexual functioning. As a patient interviewed by one team of psychiatrists reported, *"penis size is what makes you a man"* [Pope *et al*, 2002]. Furthermore a significant minority of women (~21%) attach importance to the penis size of partners [Francken *et al*, 2002].

Male dissatisfaction with penis size has spawned a whole industry dedicated to penis augmentation. Table 2 describes some of the techniques currently advertised on the internet for increasing penis size.

Table 2: Techniques for increasing penis size

Technique	Description
NON-SURGICAL	
Cosmetic	Losing abdominal fat and trimming pubic hair
Pills	Tablets and capsules typically containing herbal components purported to improve virility and penis size
Penis pump	Cylinder fits over the penis and manually or mechanically creates a vacuum which engorges the organ with blood.
Jelqing	Set of exercises that involve stretching the penis over a long period of time (often around 6 months)
Clamping	Various tools which clamp the base of the penis and so increase pressure in the corpus cavernosum
SURGICAL	
Injection	Injection of silicone or other substances (i.e. collagen or mineral oil)
Inflatable implants	The two corpus cavernosa which fill with blood during erection are surgically replaced with inflatable implants
Beading	Beads of various composition are inserted under the skin to increase girth.

These techniques are, however, associated with a number of risk factors. Penis enlargement pills have, for example, been found to contain contaminants such as mould, lead, pesticides and *E. coli* of faecal origin. Penis pumps, jelqing and clamps also risk causing physical injury to the genitals. Devices similar to penis pumps have, however, been used with some success in a clinical setting for men suffering from erectile dysfunction [Lewis & Witherington, 2005]. In America alone, some 10,000 men underwent cosmetic surgical penis enlargement in one seven year period [Wessells *et al*, 1996].

Complications have, however, included infection, scarring, bending of the penis, disfigurement and nerve damage leading to loss of sensitivity [van Driel *et al*, 1998]. Interestingly, most men seeking surgical penis enlargement actually have an averagely sized penis [Mondaini *et al*, 2002].

A penis pump.
Courtesy & Copyright Private Moments Inc.

1.1.6
MUSCULARITY

Many men in particular exhibit concerns about their level of fat and muscularity. It is thought that around 30 percent of men suffer from high levels of muscle dissatisfaction and many go to extreme lengths in order to improve their physiques [Raevuori *et al*, 2006]. These may include excessive time spent weight training, social disruption and abuse of performance enhancing drugs. At its extreme, men may be diagnosed with the body dysmorphic disorder (BDD) variant muscle dysmorphia in which men exhibit an obsessive preoccupation with their lack of muscularity [Olivardia *et al*, 2000]. This study shall explore male physique dissatisfaction as just one example of the many varied appearance issues experienced by men in the United States.

1.2
Psychosocial Causes of the Adonis Complex

The alarming increase in women suffering from eating disorders such as anorexia nervosa and bulimia nervosa has been attributed, at least in part, to television and print media. This hypothesis suggests that unrealistic media images of the female body have perpetuated a social ideal which contributes to female dissatisfaction and causes them to seek unattainable body changes. As a consequence, the argument continues, women suffer increased appearance concerns which may put them at risk of developing an eating disorder [Wolf, 1992]. Given the rising incidence of male body obsession, the relationship between this hypothesis and male eating disorders has recently drawn a lot of attention from social scientists and clinicians.

1.2.1
THE ROLE OF THE MEDIA

Media may refer to any popular means of mass communication such as television, newspapers and magazines. In the United Kingdom, around 95 percent of households possess a television, which the average person watches for between three and four hours per day. Similarly over 60 percent of men report reading a newspaper each day [Jade, 2002].

The media is known to have a profound effect on the public. Following a storyline in *Coronation Street*, for example, in which a female character died of cervical carcinoma, laboratories around the United Kingdom were inundated with requests for premature and unwarranted smear tests [Howe *et al*, 2003]. Furthermore, a vast number of surveys and studies have demonstrated an association between media images and eating disorders. One study of 232 female undergraduates, for example, showed that magazine reading was the most consistent predictor of disordered eating among these students [Harrison & Cantor, 1997]. In terms of influencing male body image, men exposed to television advertisements exhibiting ideal male body types are known to experience higher levels of body dissatisfaction than those exposed to neutral advertisements [Agliata & Tantleff-Dunn, 2004]. Furthermore, over 37 percent of men seeking surgical penis enlargement attribute their feeling of inadequacy to adolescent viewing of well-endowed men featured in pornography [Mondaini *et al*, 2002]. To quote one influential report on eating disorders: *"in a media saturated culture, the argument that long term exposure can help shape the world views of particular sections of the audience is one that merits consideration, however, the extent to which the media contribute to the personal identity remains unclear"* [British Medical Association, 2000].

1.2.2
MEDIA EXPLOITATION OF THE MALE BODY

The introduction of the male body as a sex symbol has undoubtedly increased in recent decades. A few decades ago, for example, male strip shows were confined to an almost entirely gay male audience [Pope *et al* 2002]. In the late 1970s, however, the Chippendales troupe of exotic male dancers was raised in Los Angeles and this group was soon followed by similar acts thereafter. The high degree of muscularity expected of these dancers is illustrated below.

Chippendale dancers in 2003.
Courtesy and Copyright Steve Danforth.

Although undressed bodies of both sexes are becoming increasingly commonplace in contemporary media, the male body in particular has grown exponentially in terms of its commercial value. In order to examine changing trends in media exploitation of the male body, a team of researchers surveyed advertisements printed in the major womens' magazines *Cosmopolitan* and *Glamour* over the last 40 years [Pope *et al*, 2001]. Advertisements were chosen as they are more likely to reflect the whim of consumers (i.e. women) than articles alone which may depend only on editorial policy. Although numbers of undressed female bodies remained relatively static in these publications, the number of undressed men increased significantly. This trend is illustrated by the line graph below, which has been reproduced from the original study data [Pope *et al*, 2001].

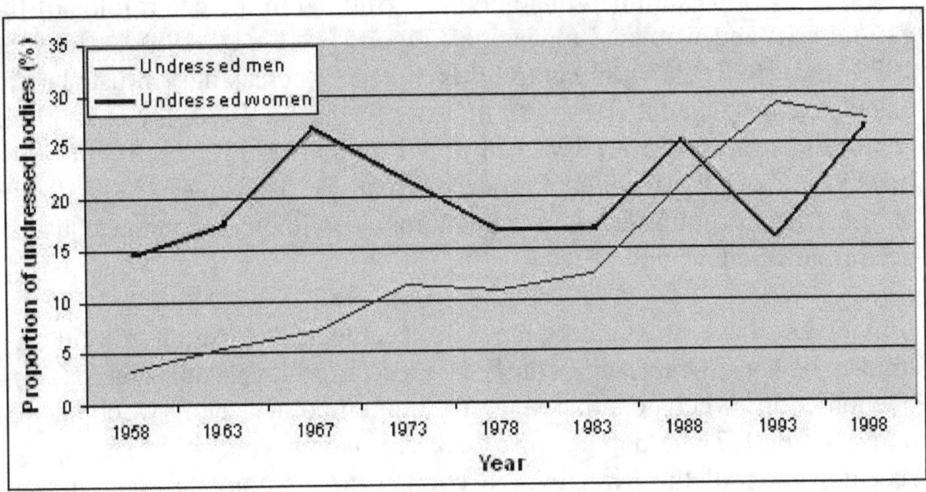

One of the most interesting observations made by this team of researchers was that undressed male bodies are not only used in advertisements for dating or appearance products. They cite numerous examples of muscular male torsos being used to sell products as diverse as ironing boards, mobile phones and bank loans [Pope *et al*, 2002]. That there exists a correlation between undressed male bodies in advertising and male body dissatisfaction has been suggested by comparisons of Taiwanese and American cultures. In Taiwan, for example, there is a lower frequency of male torso exposure in advertising and Taiwanese males are known to exhibit greater body satisfaction than American men [Yang *et al*, 2005].

1.2.3
MEDIA ASSUMPTIONS OF MUSCULARITY

Clearly the undressed male body is of greater commercial value now than it was fifty years ago. It is quite possible to image how this in itself may give rise to appearance concerns in contemporary men. The male bodies towards which modern men are exposed are not, however, only more numerous – they also increasingly boast degrees of muscularity unattainable for most ordinary men. In 1992, an ingenious study was conducted in which researchers showed that the ideal *female* body, as portrayed by *Playboy* centrefolds, became progressively thinner over a twenty year period [Andersen & Didomenico, 1992]. More recently, a team of clinicians and academics sought to examine whether the ideal media image of the male body had become more muscular. In order to assess this, they examined male bodies exhibited in the centrefolds of the gay-interest and womens' magazine *Playgirl* [Leit *et al*, 1999]. Between 1973 and 1998, *Playgirl* centrefolds became increasingly muscular, as calculated by their fat free mass index (FFMI). In 1978, for example, a centrefold model appeared with a FFMI as low as seventeen. In 1994, however, *Playgirl* exhibited a model with an FFMI of over 31 – a degree of muscularity almost certainly unattainable without recourse to use of performance enhancing drugs.

Movie stars have also undergone a noticeable shift towards muscularity. Images of Cary Grant and John Barrymore from the 1940s illustrate this phenomenon when compared with contemporary Hollywood action heroes. In 1977, the popular film *Pumping Iron* was released starring the recent winner of the Mr Olympia competition Arnold Schwarzenegger. This period similarly witnessed the launch of Sylvester Stallone in Rocky (1976) and, later, of Jean Claude Van Damme in Bloodsport (1988).

John Barrymore (1882-1942).
Courtesy and Copyright HurrellPhotos.com.

As anyone familiar with action movies will be aware, these new male role models boast physiques unattainable by all but the most dedicated of bodybuilders. In the case of Mr Schwarzenegger, it is almost certainly physiologically impossible for any of his viewers to have followed in his muscular wake. He has, for example, publicly conceded that his bodybuilding career was supported by anabolic steroids [Schwarzenegger, 2003]. As early as 1977, he wrote in a mail order pamphlet that *"anabolic steroids were helpful to me in maintaining muscle size"* [Schwarzenegger, 1977]. It is a plausible suggestion that the unobtainable muscular physiques of these action heroes had a similar effect on men as underweight catwalk models are thought to have had on young girls [Terzief, 2006].

1.2.4
THE EVOLVING MUSCULATURE OF MALE ACTION TOYS

Indeed, the commercial use of muscular role models is not limited to print media. Unattainable levels of muscularity are also found across a number of action figures typically bought for young boys. This phenomenon was first noted in 1995 when researchers compared the measurements of Ken and Barbie dolls to those of healthy adults [Brownell & Napolitano, 1995]. These workers found that, in order to achieve Barbie or Ken proportions;

1. A healthy female would need to grow 24" while losing 6" from her waist.
2. A healthy male would need to grow 24", add 11" to his chest and 7.9" to his neck.

Researchers interested in male body dissatisfaction have taken this observation further by studying the evolution of male action figures. A collection of various GI Joe and Star Wars characters was first amassed with representatives from different years over the last three decades. These were then measured and scaled to a common height of 1.78m [Pope *et al*, 1999].

These workers found that, even before attempting to quantify muscularity, it was clear that later models of GI Joe were more muscular than their earlier counterparts. GI Joe Land Adventurer, for example, had no visible abdominal muscles in 1964 whereas, by 1994, these were clearly defined. They also showed that the 1998 GI Joe Extreme Sergeant Savage model would, if scaled to 1.78m, possess a degree of muscularity only ever achieved by the most serious of competition bodybuilders. To illustrate this finding, GI Joe Sgt. Savage is shown alongside the earlier GI Joe Land Adventurer model.

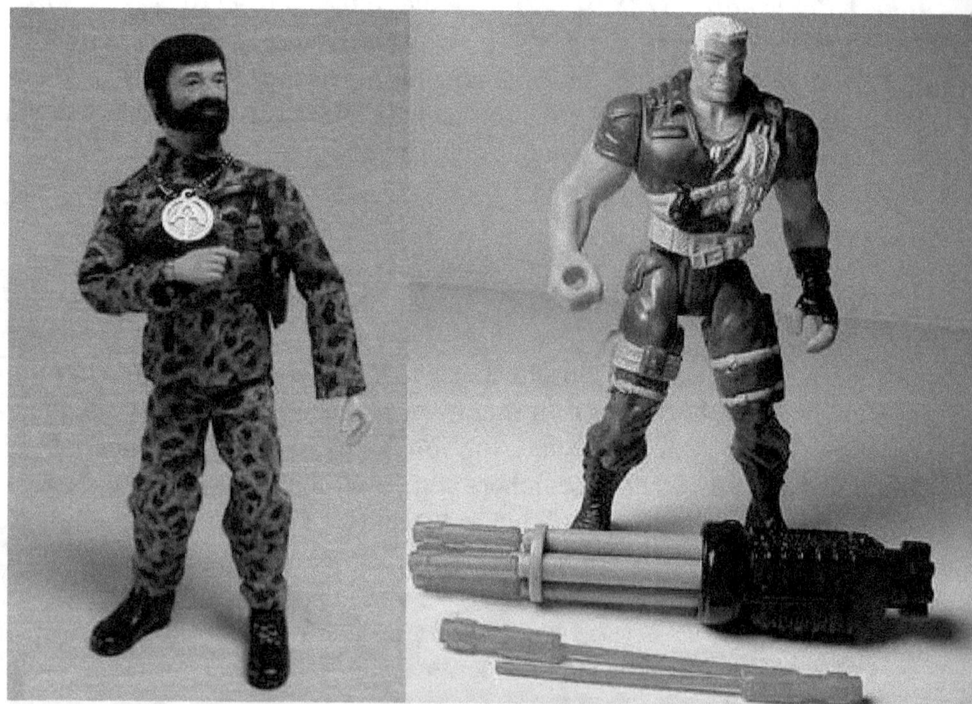

GI Joe Land Adventurer from the mid-60s (left) and GI Joe Extreme Sgt. Savage from the 1990s.
Courtesy and Copyright Brian Savage (left) and Randy Herkovitz (right).

Table 1: GI Joe Extreme Sergeant Savage takes on Arnold "Mr Universe" Schwarzenegger		
	GI Joe Extreme Sergeant Savage	Arnold Schwarzenegger
Chest	54.8"	57"
Bicep	26.8"	22"
Waist	36.5"	34"

The unachievable muscularity of Sgt. Savage is illustrated by Table 1 which compares his proportions to those attributed to Arnold Schwarzenegger at the peak of his bodybuilding career [(Pope *et al*, 1999) & (Schwarzenegger, 2003)]. Mr Schwarzenegger is considered to be an icon among contemporary bodybuilders, winning eight Mr Olympia and seven Mr Universe titles [Schwarzenegger, 2003]. As has already been noted, Mr Schwarzenegger has since conceded long term use of anabolic steroids to achieve these sizes [(Hartlaub, 2003) & (Schwarzenegger, 1977].

1.2.5
A SOCIAL EXPLANATION FOR THE ADONIS COMPLEX

A number of social scientists have attempted to explain the media shift which has undoubtedly contributed in some way to the male pursuit of muscle evident today. They point in particular to the fact that men are now increasingly relinquishing their traditional masculine roles to women. Indeed, women continue to join traditionally male professions, command high salaries and are more likely to be independent of men than ever before. These scholars claim that, as a consequence, the male body has assumed increasing importance as the one feature which continues to define men as masculine. As the male role of breadwinner has declined, the value of his physical appearance has risen [Gillet & White, 1992]. This is likely to have been compounded by the increased female choice in mate selection. Indeed, women who are more attractive, educated and financially secure are most likely to attach importance to the physique of potential partners [Neimark, 1994]. In order to further support this hypothesis, a number of scholars have produced a timeline comparing feminist advances and events associated with male body insecurities [Pope *et al*, 2002]. Although this crude comparison cannot be considered as evidence – it does not even attempt to eliminate confounding factors – it does add an interesting flavour to the hypothesis. This timeline has been reproduced below:

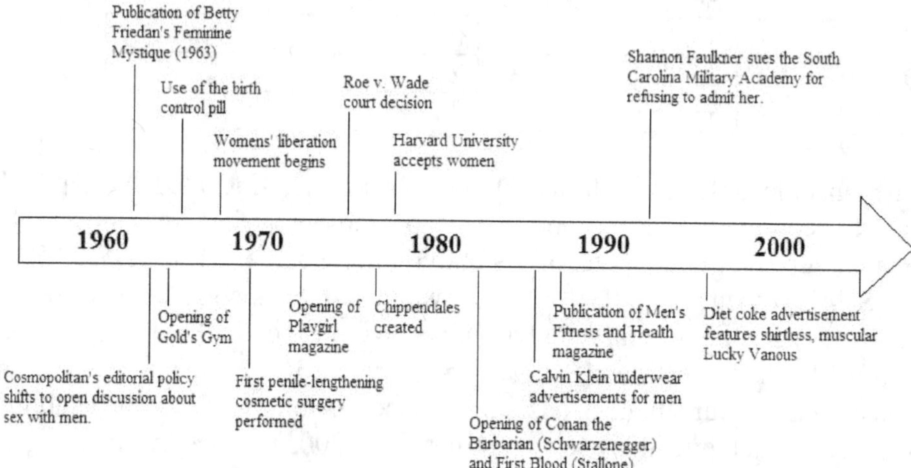

A timeline indicating major feminist advances (above) and landmarks in male body image (below).
Figure reproduced from Pope et al, 2002.

In further support of this hypothesis are the results of studies in cultures which have retained traditional gender roles. One team of researchers, for example, has found that Taiwanese men exhibit significantly less body dissatisfaction than their Western counterparts [Yang *et al*, 2005].

1.2.6
CONCLUSION

From the examples and studies described above, it is clear that, not only is the male body now used more frequently for commercial purposes, but the models used are becoming increasingly muscular. Young boys are now exposed, from an early age, to toys which exhibit hypermasculine physiques. As they grow older they come to treat muscular actors such as Sylvester Stallone, Arnold Schwarzenegger and Jean Claude Van Damme as role models. Many young American men further develop an interest in WWF Wrestling and the massive physiques associated with this mode of entertainment. Perhaps more significantly, they are bombarded by magazines, advertisements and television with countless media images reminding them of the 'ideal' male physique. The men used in these images have often reached proportions impossible to achieve without years of training and, increasingly, without abuse of performance enhancing drugs.

Despite these pressures, of course, many men are able to continue leading normal, healthy and balanced lifestyles. Many others are, however, more vulnerable to these media messages and some may already have an innate predisposition to be insecure about their body shape. It is perhaps little wonder that male appearance concerns have spawned a multi billion dollar industry and that psychiatrists are recording an unprecedented rise in psychiatric appearance pathologies among men.

1.3

Sociobiological Causes of the Adonis Complex

Although most research into male body image has approached the subject from a psychosocial angle, this author is naturally suspicious of explanations which fail to give due consideration to evolutionary causes. The following section explores the relationship between social and evolutionary explanations for male body dissatisfaction and, in particular, male pursuit of increasing muscularity.

1.3.1

THE ADONIS COMPLEX AS A UNIVERSAL PHENOMENON

On a research project in Malawi last year, this author made the observation that part of the stigmatisation of men with HIV/AIDS was the accompanying muscle wastage and associated *"weakness"*. Despite the absence of weight training facilities, young Malawian men did look up to others who were more muscular and so considered to be *"strong"*. This suggests that the male pursuit of muscularity is not limited solely to Western ideals of masculinity.

Fortunately a number of studies have examined body image perceptions in non-Western men. One group of researchers, for example, studied the Ariaal of Northern Kenya who are pastoral nomads and so have little contact with Western print or film media [Campbell *et al*, 2005]. Indeed, ethnographic observations suggested that, even in the nearest town, there was only one television. Study participants were asked to choose from a number of different images which a) represented their own body, b) the body of an average man in their region, c) his ideal body and d) the body type to which he thought women would be most attracted. Interestingly, Ariaal men under-judged their level of muscularity compared to the 'average male body' by far more than did their American counterparts. As these researchers concluded, *"that a preference for greater muscularity exists among men even in a culture where chronic energy deficiency makes such muscularity difficult to attain suggests that muscularity is widely valued as an ideal male trait"*. An almost identical study carried out in young Samoan men served to validate this conclusion. Indeed, these South Pacific islanders chose an ideal body type 3-5kg leaner and 8-11kg more muscular than their own [Lipinski & Pope, 2002].

Further evidence against the view that the male pursuit for muscularity is a socially constructed phenomenon comes from examination of historical evidence. Although paintings such as Titan's Venus and Adonis suggests that male beauty was not historically dependent on muscularity, statues from classical Greece indicate that muscular physiques have always been associated with power and authority.

A statue of Ancient Roman origin.
Courtesy & Copyright Chatsworth Settlement Trustees.

A ceiling painting in the Drawing Room of Chatsworth House, Home of the Dukes of Devonshire. The painting, by Antonio Verrio (1639-1707), depicts a number of muscular male characters.
Courtesy & Copyright Chatsworth House Trustees

1.3.2
THE EVOLUTIONARY BASIS OF LARGE MALE BODY SIZE

Western culture idealises the large, muscular man and the small slim woman. Although exceptions occur, particularly in anorexic men and bodybuilding women, men will therefore seek to gain mass in terms of muscle whereas women prefer to lose mass in terms of fat. This is interesting as, in physiological terms; males tend to be naturally larger than females. The average Caucasian American male for example reaches 178.2cm and weighs in at 191lbs compared to the average female at only 164.1cm and 164lbs [Ogden *et al*, 2004]. Interestingly, this observation that males are typically greater in size and weight holds true across almost all animal species [Wilson, 2000]. That this trait has evolved across such a range of organisms suggests that male 'largeness' might be a valuable trait in terms of survival and reproductive potential. Indeed, among polygynous mammals, the ratio of the male's body size to that of the female increases with the size of the average harem controlled by males of that species [Diamond, 1992]. Table 1 illustrates this principle using a number of mammalian examples. This difference in body size between males and females is known to zoologists as sexual dimorphism. Along with this phenomenon comes the development of increasingly obvious secondary sexual characteristics – the external features indicative of gender which are separate from the reproductive organs themselves. Sexual dimorphism, including body size and secondary sexual characteristics, are known to become more pronounced in naturally polygynous species [Wilson, 2000]. In humans, which are considered by anthropologists to be mildly polygynous, secondary sexual characteristics include breast shape, facial hair, vocal tone, body size and musculature [Dixson *et al*, 2005].

Table 1: Mating habits and male to female body size ratio		
Mammalian example	**Mating habits**	**Male/female body size ratio**
Gibbon *Hylobates hoolock*	Monogamous	Difference in body weight not statistically significant. Both ~7kg [Geissmann, 1998]
Gorilla *Gorilla gorilla*	Harem size of 3-6 females	Males average 181kg compared to females at 85kg [Long, 2006]
Southern elephant seal *Mirounga leonina*	Harem size of up to 40 females. Only 3% males breed in each season	Males average 2045kg compared to females at 680kg. Males may weigh more than 10 times that of reproducing females [McCann *et al*, 1989]

The evolutionary rationale for this difference in body size between genders is that, as mating becomes competitive, natural selection drives greater morphological changes. In terms of body size, then, males will tend to become progressively larger in order to dominate females and to fight or otherwise intimidate other males [Wilson 2000]. To put this another way, if large males are more likely to produce more offspring, genes for 'large male body size' will become increasingly prevalent in the population. This effect may be compounded by further sexual selection, i.e. females will become more attracted to large males as their offspring will tend towards more propagation if allied to genes for male largeness. The combined result is increasing body size among males. This concept has been tested in other organisms. In one lizard species, for example, large head size has been associated with greater combative ability among males and an associated increase in the number of sexual encounters. Head size in males of this species has also been shown to correlate historically with the need for male combat in order to win the attention of mates [Kratochvil & Frynta, 2002]. This explanation for greater male body size may also explain the predisposition of males to seek greater size through weight training.

1.2.3
CONCLUSION

The evolutionary evidence available suggests that, in physiological terms, males become larger as a response to competition for mates. This explanation is reminiscent of the sociological suggestion that the male pursuit of muscularity is related to contemporary changes in gender roles. As the economic position of women improves, they consequently acquire a better position to pick and choose mates for reasons over and above job status and a comfortable salary. In the short-term, one likely consequence of this transition is increased media exploitation of the male body (i.e. to a female market) and a consequent explosion in male body dissatisfaction. In the long term, evolutionary biologists might expect to observe increased sexual dimorphism as male genes 'compete' with one another to produce bodies which attract more sexual encounters. This hypothesis may provide a site of convergence for both evolutionary and social explanations for the Adonis Complex.

1.4

Why Men Weight Train

In order to determine why men weight train, bodybuilders leaving Muscle Beach were asked for their comments on why they thought they continue to lift weights. These explanations were varied but could broadly be categorised into three classes, which were used to inform the design of a questionnaire. This questionnaire was returned by 228 men, each of whom trained at least three times a week and exhibited a higher than average degree of muscularity. The categories of explanation used were "attracting potential partners", "achieving status among peers" and "improving health or ability for a particular sport". Respondents were asked to assign a percentage to each factor in order to indicate its relative importance in their decision to lift weights.

1.4.1

ATTRACTING POTENTIAL PARTNERS

In total, respondents assigned 34.3 percent to 'attracting potential partners' as their reason for weight training. One 35 year-old social worker explained that he was motivated by making *"the chicks drool and lick their lips when [they] walk past [him]"*. Indeed, if many of these men are to be believed, their efforts in the gym are paying off in terms of sexual interest. One said that he was *"getting a lot more looks [his] way"*, another that *"the girls dig abs…there is no two ways about it"* and a third that *"the bottom line is that bodybuilding equals more looks from girls"*. Some men knew their muscularity to be a source of female attention because they were often asked to flex biceps or show off their abdominal muscles. Others, however, thought that weight training had led to more female interest because it had increased their own confidence. *"When I was skinny"*, said one seventeen year-old, *"I felt more insecure around girls…now that I'm heavier I feel better for some reason"*.

Some men, however, disagreed that their muscularity had led to increased success in attracting partners. One of the most surprising responses came from a nineteen year-old who said that, although he attracts more attention from girls, *"most of my time is spent worrying about what I eat, exercising, and working out...so I haven't really been trying to get with any girls*". Other men betrayed low self worth with observations such as "*you can have a great body but if your face isn't up to it...*" and "*girls also care about personality... so I'm ****ed*".

Others believed that women were not attracted to men with too much muscle. Men were divided over whether they considered this to be a limitation on the degree of muscularity they hoped to reach. A number, for example, agreed with one man who was anxious not to grow so large that he looked "*distorted*". Others, however, wanted to be 'huge' regardless of whether this limited their success with the opposite sex. For example, one nineteen year-old described how, when he flexes his bicep, "*girls are just grossed out*". Another purposefully wore long sleeved shirts in public so that girls were not disgusted by the veins and muscles in his arms. A third man described how his "*girl friends say they don't like muscle*" but added "*whatever... I don't care... I'm still gonna carry on getting built and ripped*".

In all, these responses suggested the existence of two distinct sub-populations. The first of these were motivated, at least in part, by their perception that potential partners would be more interested in a muscular physique. Some of these men had high self esteem and sought to maintain or improve their level of success. Others appeared to suffer from low self-worth and considered bodybuilding as a way of compensating for other perceived defects in their appearance. The second sub-population, however, appeared to disregard the views of potential partners in terms of whether they might be considered to be too big. These men were often already, or had ambitions to one day become, professional or semi-professional bodybuilders. Indeed, these bodybuilders often reacted with distain to the suggestion that they might be motivated by sexual interest. As one man said, echoing a sentiment expressed by many others, "*if a woman wants a small skinny guy then she is not the one for me*".

1.4.2
ACHIEVING STATUS AMONG PEERS

Respondents assigned 29.1 percent to 'achieving status among peers' as their explanation for bodybuilding. Most explained this in terms of increased confidence, which they felt gave them more presence and impact in social situations. This, they said, had implications for how they were treated by their peers but also by teachers, employers and others to whom they are brought into contact each day. Although one respondent thought that some *"skinny people can get a lot of attention just walking into a room"*, others felt that they were *"definitely noticed more"* and *"got a load more friends"* since starting weight training. The consensus was summed up by one Australian visitor to LA who thought that *"being built is definitely socially acceptable... the results of bodybuilding spill over into everything... and almost always in a good way"*.

For a number of men, however, the status they were hoping to achieve was limited simply to other men in the gym. One man in particular spoke for many when he expressed his greatest motivation as a desire to be *"the biggest, baddest sonofa***** in my gym"*.

1.4.3
IMPROVING HEALTH OR ABILITY FOR A PARTICULAR
SPORT

This final category was the most popular choice among respondents by a small margin. In total, respondents assigned 36.6 percent of their reasons for weight training to 'improving health or ability for a particular sport'. The most popular sports associated with weight training were football (of the American variety) and wrestling. Indeed, a significant number of men explained that playing these sports at high school had first introduced them to bodybuilding. Others in this category, however, weight trained simply for the sake of bodybuilding itself. Fifteen percent of questionnaire respondents, for example, were already, or had ambitions to become, a competitive bodybuilder.

Many, however, felt that bodybuilding improved their health. One man told me that that morning over breakfast; his wife had asked him to stop using the popular bodybuilding supplement creatine monophosphate. *"What right does she have"* he asked *"to lecture me about health when she is sat there eating cereal packed with sugar, salt, fat... and all sorts of artificial colours and flavourings?"*. Similarly a number of college students claimed that they had become more focussed academically since starting weight training because they felt generally more alert and healthy. One twenty year-old explained how he had stopped drinking, smoking and eating unhealthily as he was now motivated by his weight training to stay healthy.

1.5
What Do Women Think?

As has been shown in previous sections, men harbour a variety of appearance concerns ranging from facial attractiveness to genital size to physique. In many cases men explained these concerns in terms of how their perceived defects might affect their relationships with women. Indeed, one survey discussed previously showed that bodybuilding men assigned 34.3 percent of their explanation for weight training behaviour to wanting to attract potential partners. Furthermore, body dysmorphic men often use the opposite sex as partial explanation for their obsessive appearance concerns. One man, for example, wrote that *"girls like tall muscular men with blue or green eyes, a big ass, and a manly face. Why did I get cursed like this?"*. Another described the onset of his illness as following a conversation with a female friend about her attraction to a *"hot guy"* she had met online. *"Then it was like my eyes were opened to a very sad truth about the world..."* he wrote *"...I dunno why but this just affected and depressed me greatly... from then on I was envious of anyone that was good looking..."* This section, then, will consider the degree of muscularity considered to be ideal in a man by women and how this relates to what men believe women consider to be ideal.

1.5.1
PREVIOUS WORK ON THE FEMALE PERSPECTIVE

A number of studies have shown that men consistently over-estimate what features are considered to be ideal by the opposite sex [Pope *et al*, 2002]. Around a decade ago, two researchers measured the physical attributes of white heterosexual students in order to explore perceived and actual ideals held about each sex. These workers found that both men and women exaggerated the features which the opposite sex idealised about their own [Jacobi & Cash, 1994]. Another study using line drawings has shown that men believe that women seek a more muscular stature than they actually do [Fallon & Rozin, 1985]. More recently, a survey of romantically involved students showed that, despite being familiar with one another, men and women over-estimated their partner's ideals by a strongly comparable degree [Tantleff-Dunn & Thompson, 1995]. In terms of muscularity, one study of men in Austria, France and the USA showed that men across all three countries chose a body 30lbs more muscular that their own when asked which women would consider to be ideal [Pope *et al*, 2000]. Interestingly a similar study conducted in middle-aged men failed to show this discrepancy which suggests that men in later generations are more affected by the Adonis Complex [Lynch & Zellner, 1999]. Clearly, then, women on the whole prefer a much lower level of muscularity than men of the same age believe that they would.

1.5.2

A QUALITATIVE EXPLORATION OF THE FEMALE PERSPECTIVE

In order to assess how women respond to very muscular men, a small sample (n=8) were asked to comment independently of one another on six images of shirtless men. The men in these images ranged from slim through lean to muscular and ultimately to competition bodybuilder standard. Unfortunately it has not proved possible to reproduce the images used in this survey for copyright reasons.

All of the women asked immediately disregarded the possibility that the competitive bodybuilder might be attractive. This validates previous research which found that, when women were asked to rate competitive bodybuilders on a scale of one to five, 94 percent ranked these men as 'five' or *"extremely repulsive"* [Klein, 1990]. Equally, 88 percent of the female participants disregarded the slimmest man, variously describing him as *"looking weak"*, *"small"* and *"petite"*. One girl further commented that she would worry about weighing more than this man. After eliminating the largest and smallest of the men shown, the women were then divided almost equally between whether they favoured a very lean or a muscular physique. 38 percent (3/8) liked what they called the *"bodybuilder"* or *"Van Damme"* look and what one described as *"washboard abs, chiselled arms, and sculpted shoulders"*. One commented that she likes *"[her] men strong"*, another that *"my man must be able to pick me up without straining or he's out"* while a third said that *"tall muscular supermen are great for casual affairs...but I wouldn't trust someone that hot in a long term relationship... for [that] I prefer someone less perfect, who worships me... and doesn't have a dozen other females sniffing around"*. Of the remaining 62 percent (5/8), 38 percent (3/8) did not like much muscle on men – one going so far as to describe the muscular male images as *"nasty"*. Another said that *"the overmuscled look is gross... prettyboys and bodybuilders don't interest me at all"*. All three echoed the sentiment that many men grow far too large in their pursuit of muscle and that their physique becomes distorted beyond what these women could consider attractive. One further associated high muscularity with vanity which she considered to be a repulsive feature in potential partners. In her words, *"when I was younger I liked that sort of physique... but now I know they must spend so much time working to stay that way... he probably wouldn't have much time for me"*.

Perhaps surprisingly, 25 percent (2/8) of those asked favoured a greater level of body fat on potential partners. These they called *"cute"* and *"teddy bear types"*. One woman described her preference as *"normal looking men... with a beer belly, hair, broad shoulders... tall"* while another chose a *"normal, cuddly guy"*. Another did not like any of the images presented to her and instead preferred men *"with a little more body fat"*.

1.5.3
CONCLUSION

These responses suggest that there is great diversity of opinion among women as to what constitutes the ideal male physique. Despite this diversity, it is clear that women do not tend to favour either very thin or very muscular men. Once in between these extremes, however, opinion is strongly divided over the optimum size, muscularity and fat composition of the male body. This is particularly interesting given the value of 34.3 percent which bodybuilding respondents assigned to 'attracting potential partners' as their explanation for weight training.

1.6

Summary

1. Contemporary men are more concerned about their appearance than were their predecessors of decades ago. Areas of greatest concern are height, facial appearance, skin, hair thinning, genitals and general muscularity.

2. Males are larger than females for evolutionary reasons and this is reflected by the preference of men to train for greater size whereas women more frequently train to become smaller.

3. A culture shift has occurred in recent years so that more attention is now focussed on the male body as an object of desire and success. This is, in part, compounded by changing attitudes reflected in print and film media. It may also be a consequence of the women's liberation movement which has given women freedom to expect more than a large salary from potential mates.

4. Bodybuilding men explain their weight training in terms of improving health, gaining more status and of attracting potential partners.

5. Despite this, it would seem that men over-estimate the aesthetic expectations of women. There is no clear agreement between women on what constitutes the ideal male physique or, indeed, on how important this element is in terms of selecting partners.

PART II

The Rise of Male Body Obsession

All that glisters is not gold;
Often have you heard that told:
Many a man his life hath sold
But my outside to behold:
Gilded tombs do worms enfold.

Shakespeare, The Merchant of Venice (II, viii)

Part II

2.1

Social and Physical Dangers of Bodybuilding

Successful accumulation of lean muscle mass requires commitment to a number of principles. These include, of course, weight lifting but also optimum patterns of eating, sleeping and use of dietary supplements. To many outsiders, strong adherence to these principles may appear obsessive. In psychiatric terms, obsession describes a persistent compulsion which generally impedes normal function and/or causes distress (Funk, 2006). On the whole, however, bodybuilders strongly rejected this interpretation of their behaviour. Indeed, one bodybuilder summarised the general response to these critics with the words *"obsessed is just a word the lazy use to describe the committed"*. In order to resolve this conflict, a questionnaire was administered to 228 bodybuilding men enquiring about the time, money and energy invested in their approach to bodybuilding. Questionnaire responses were analysed, and the results used to provoke comment in interviews with other bodybuilders.

2.1.1
WEIGHT LIFTING

Bodybuilders lift weights in order to challenge muscle fibres. The intention is to lift weights heavy enough to stress the muscles which will encourage them to grow larger so as to resist similar strains in the future. Physiologically, this process is known as muscle hypertrophy. While most interviewees belonged to gyms, a small number had collected enough weight lifting apparatus to have installed their own home gyms. On the whole, interviewees preferred to lift weights in gyms because of the professional equipment available, music and motivating atmosphere.

Reported time spent lifting weights per week ranged from between 2 and 30 hours with an overall average of 7.2 hours/week. Although it is difficult to distinguish between normal and obsessive weight lifting from a simple quantitative assessment of time spent in the gym, 30 hours per week is clearly excessive. Indeed, this amounts to over four hours spent working out *every* day. Interview responses further support these findings. One eighteen year-old, for example, said that "*I think about bodybuilding all day... I can't even go to sleep cause I think about how I'm gonna lift the next day*". Others also exhibited a compulsive need to weight train. A younger teenager described the feeling after missing a workout as "*going crazy... it feels like my world is falling down*" while another said that "*nothing ever keeps [him] from lifting*".

Interestingly men reporting over fifteen hours/week spent weight lifting were more likely to explain their behaviour in terms of 'gaining status among peers' or 'attracting potential partners'. This is in contrast to those lifting weights for less time who preferred to explain their efforts in terms of improving health or increasing sporting ability. This suggests that appearance concerns are more likely to provoke a higher level commitment to bodybuilding behaviours than are health or sporting reasons.

In order to determine the impact of bodybuilding on the lives of respondents, questionnaires asked whether weight lifting ever prevented socialising with friends. 53 percent (116/217) agreed that weight lifting does sometimes prevent socialising with friends. Interviewees explained this in terms of the rigorous self-imposed weight lifting schedule and a perceived need to avoid alcohol. Perhaps unsurprisingly, a higher proportion – 83 percent (5/6) – of men reporting over 15 hours weight training per week agreed that this activity interrupts their social lives.

Besides interfering with day-to-day life, excessive weight lifting may also cause physiological damage to muscles, joints or tendons. Orthopaedic and radiological examination of bodybuilders has shown that the shoulder and elbow joins as well as the lower back and knees are most at risk [Goertzen *et al*, 1989]. Muscular trauma such as pulls, sprains and tendonitis accounted for 83 percent of injuries reported by respondents. Many of the men interviewed did not use a proper warm up routine and this may underlie some of the musculoskeletal injuries sustained in the course of weight training. One eighteen year-old for example told how he often lifts weights so heavy that he feels the need to vomit during work outs. In an attempt to explain this statement, his training partner observed that "*pain is like... weakness leaving the body*".

2.1.2
EATING

Once muscle fibres have been challenged by resistance weight training, their repair is necessarily facilitated by maintenance of a high quality diet. Growth is permitted by eating a high-calorie, high-protein diet. Bodybuilders attempt to ingest as many as 6000 calories a day and around two grams of protein per kilogram of body weight. This is twice the recommended daily calorie intake and more than twice the recommended daily allowance (RDA) of protein for a young man [Boothby, 2004]. In order to consume this volume, interviewees ate as many as six meals and often, additionally, used purified sources of protein which are mixed into a smoothie or milk shake.

A number of men echoed the phrase *"eat big to get big"* and many cited eating too little food as their biggest obstacle to accruing lean muscle mass. Respondents reported spending between 1.5 and 25 hours a week buying food, preparing meals and eating to satisfy their special dietary requirements. On average, respondents spent seven hours/week on these activities in addition to the time spent actually lifting weights. The total time spent fulfilling lifting and eating requirements then totals an average of 14.2 hours per week – over two hours every day – for the average male weight lifter. As a high protein diet necessarily involves expensive meat products as well as food supplements, respondents reported spending as much as $300 (£160) per week on food, dietary supplements and gym memberships.

Underlying these figures were some rather alarming interview responses. One eighteen year-old expressed himself as follows: *"I eat until I feel sick then eat some more.. it makes me feel good about myself and I know it helps me grow"*. This statement alone almost satisfies the diagnostic criteria for binge eating disorder [American Psychiatric Association, 2000]. Many men additionally suggested that their social lives, personal relationships and even careers were disrupted by a need to eat every three hours. One man, for example, took work as a personal trainer simply because a regular nine-to-five job would not allow him to continue his eating schedule. Similarly a high school student found that *"in school [he is] always thinking that [he needs] to eat 185g of protein"*. Many men described feeling frustrated and angry if prevented from eating for longer than they perceived was necessary to preserve optimum muscle mass. As one man reported, *"when I get over three hours of eating because something happens... well... it always ****es me off and I have to find a way to eat no matter what"*. The girlfriend of another bodybuilder talked to me while her boyfriend trained on Muscle Beach. She described the temper tantrums which occur frequently when her boyfriend is unable to eat regularly enough.

2.1.3
SLEEPING

Many men cited sleep as the third key principle of bodybuilding. Interviewees explained that, during sleep, muscles repair and growth hormone is secreted to promote muscle growth. Lack of sleep, on the other hand results in increased cortisol production which supports breakdown of muscle. Indeed, these beliefs find considerable support in the medical literature [Cauter *et al*, 1997].

Once again, however, this requirement caused significant disruption to the lives of many men. Of the 53 percent (116/217) whose social lives were disrupted by bodybuilding, many cited a need to maintain optimum sleep patterns as their primary reason to avoid socialising. *"I can actually see that I am smaller..."*, reported one respondent, *"...when I have not slept a full eight hours the night before... it's demoralising to see that loss when it's all so hard to gain"*. Another man claimed that he feels irritable the following day if he failed to sleep for eight consecutive hours *"not because I am physically feeling like **** but because I missed my target"*.

2.1.4
USE OF DIETARY SUPPLEMENTS

Over 80 percent of those surveyed had used over-the-counter chemical or dietary supplements in order to improve their muscle gains. The substances reportedly used by those surveyed have been included in Table 1. By far the most commonly reported products were artificial sources of protein such as bars and shakes, presumably to satisfy the dietary requirement of two grams of protein per kilogram of bodyweight per day. Sources of purified whey protein sold in the GMC health shop in nearby Santa Monica contained around seventeen grams of protein per scoop. This is typically dissolved in milk, water or fruit juice and these products are sold in a variety of flavours such as strawberry, vanilla and chocolate. Indeed, protein supplementation is so acceptable in Los Angeles that most cafes which sell fruit smoothies will add a scoop of whey protein for around $1 extra.   Although these products are unlikely to cause significant health problems, they may act as an introduction for men to the use of other, potentially more dangerous, dietary supplements aimed at improving musculature.

Creatine monophosphate is another commonly used over-the-counter dietary supplement which was reportedly consumed by 45 percent (102/228) of those surveyed. Bodybuilders typically 'load' using one heaped teaspoon (25g) of creatine five times a day for a week before changing to a smaller dose. The purpose of this product is to increase phosphocreatine synthesis and so to increase isokinetic torque and strength. Although use of creatine at these levels is thought to have little negative effect in healthy individuals, it has been associated with cramps, dehydration, diarrhoea and dizziness [Graham & Hatton, 1999]. Men with a history of renal disease or taking nephrotoxic medications may additionally be at risk of renal dysfunction when using creatine [Yoshizumi & Tsourounis, 2004]. Furthermore, the long term effects of high doses of creatine, even in healthy individuals, remain unclear [Juhn & Tarnopolsky, 1998].

Table 1: Over-the-counter dietary supplement and drug use

Product	Prevalence of use
Artificial sources of amino acids	36% (81/228)
Creatine monophosphate	45% (102/228)
Ephedrine	3% (7/228)
Natural growth hormone boosters	4.4% (10/228)
Natural testosterone boosters	10.1% (23/228)
Nitric oxide	<0.01% (3/228)
Protein bars and shakes	80.3% (183/228)

Perhaps the most dangerous supplement used by interviewees and survey respondents is ephedrine. This product works as an appetite suppressant but is actually very similar in structure to its synthetic derivatives amphetamine and methamphetamine. Fat loss supplements containing ephedrine may result in dependence or, more seriously, in a whole host of cardiovascular complications including mortality [Berman *et al*, 2006]. Ephedrine has, however, recently been prohibited by its addition to an annex of an item of international law [INCB, 2006]. As a consequence it is likely that the usage prevalence reported by these respondents is likely to fall in the near future.

2.1.5
ADDITIONAL APPEARANCE RELATED ACTIVITIES

55.4 percent (124/224) of respondents claimed not to spend any extra time on their appearance over and above an expected level of shaving and showering. Others, however, spent as many as eight hours a week on activities such as modifying their hair and tanning. Perhaps unsurprisingly the men reporting more than five hours a week on these additional activities were much more likely to explain their weight lifting behaviour in cosmetic terms than improving health. This finding again suggests the existence of two distinct sub-populations of weightlifters, one of which hopes to improve appearance and the other of which sincerely hopes to maximise health.

2.1.6
PREOCCUPATION WITH MUSCULARITY AND APPEARANCE

The high level of commitment shown by some men to their bodybuilding rituals may, however, betray more about the nature of bodybuilding than of the bodybuilders themselves. There is a strong consensus in the bodybuilding community that lean muscle mass can only be achieved by hard work and dedication. As one interviewee reflected, every sport has athletes who dedicate huge amounts of time to their diet and training regimen. Without these people we would not have sports stars and yet we do not often level the charge of 'obsession' at them. This explanation is not, however, sufficient to enable us to overlook some of the more alarming interview responses. One man for example admitted that he would alternate hands between strokes when shaving so as not to risk one bicep growing larger than the other. This preoccupation with his physique clearly extends beyond what might be considered normal behaviour, even within the context of a demanding sport.

In order to provide a more objective measure of the seriousness of this commitment, an attempt was made to assess the importance attached to bodybuilding relative to another significant entity – life itself. To achieve this, respondents were asked whether or not they would sacrifice three years of their life to possess their ideal body. A worrying 63.1 percent (137/217) responded in the affirmative. An online focus group was then held with seven weight lifting teenagers to explore this phenomenon in more detail. In this discussion, one nineteen year-old claimed that he would happily sacrifice as many as twenty years of his life. *"The ideal body helps every aspect of your life, and the closer you get to it the easier life gets"*, he wrote, *"...the 30 years of living life with the flawless body would be so much better than the 50 years of living life with a s****y body"*. Another nineteen year-old respondent described this comment as the *"vainest thing [he] had ever heard"* and then went on to write that he could only imagine sacrificing two years. Others however disagreed with the whole idea of sacrificing life for appearance. An eighteen year-old wrote that he would *"rather be a skinny happy guy than a miserable jacked guy... no-one should be sacrificing the joys of life for a better body"*. Others felt that time spent in the gym was not a sacrifice of time but an investment and, indeed, often something to be enjoyed. On the whole, opinion was split quite evenly over the principle of whether muscularity was worth the sacrifice of years of life. Once again, this suggests the existence of two distinct sub-populations in terms of priorities and the lengths to which men might go to improve their physiques.

2.1.7

CONCLUSION

Although it is impossible to state in absolute terms what might constitute obsessive behaviour, some of the data and many of the interview responses clearly indicate an unhealthy preoccupation with bodybuilding rituals. Interestingly, the same respondents tended to report an excessive amount of time weight lifting, adherence to disciplined eating regimes and use of over-the-counter supplements. This suggests that a distinct sub-population exists among those surveyed. On closer examination of these questionnaire responses, it became apparent that these men account for a high proportion of those motivated by a desire to attract potential partners and improve social status.

This indicates that it might not be possible, after all, to determine to what extent bodybuilders as a group are committed or obsessed about their 'hobby'. The reality is that there are at least two different classes of bodybuilder with their own motivations, preoccupations and risk factors for health. Of particular concern are the group which might be dubbed the cosmetic bodybuilders. These men appear are more likely to expose themselves to musculoskeletal damage in the gym and to sacrifice their social and work lives for their appearance. They are also more likely to agree to sacrificing years of their life in order to gain muscularity. This final question was not, however, asked in a purely hypothetical and abstract manner – it does have genuine implications for real life. As will become apparent in the next section, these men are also more likely to risk their health by abusing illegal performance enhancing drugs.

2.2
Abuse of Anabolic-Androgenic Steroids

Anabolic-androgenic steroids (AAS) are a class of natural and synthetic steroid hormones which promote tissue growth in the body including that of muscle. For this reason, competition bodybuilders are known to use illegal AAS products in order to gain a chemical advantage over their competitors. Within the last fifteen years, however, AAS have become increasingly popular among high school sports players and amateur weight lifters.

In order to characterize trends among AAS users, an anonymous online questionnaire was marketed at a number of websites and among serious bodybuilders on Muscle Beach in Los Angeles. Of the 91 AAS users who completed the survey, 8.9% (8/90) were competition bodybuilders, 25.6% (23/90) hoped to become a competition bodybuilder and 65.6% (59/90) had no such ambitions. All respondents were male and most, over 75 percent, hailed from the United States. They ranged from sixteen to 44 years of age with an average age of 25. Respondents were asked only to complete the questionnaire if they had *"completed one or more full cycles of an anabolic steroid not prescribed by a doctor"*.

2.2.1
AAS AND COMPETITIVE BODYBUILDING

A number of commentators noted in the 1950s that Soviet athletes were dominating international strength events. When it became apparent that this domination was a result of regular testosterone injections, a team of physicians led by John Ziegler sought to develop an oral equivalent for use by American athletes. As a result, the federal Food and Drugs Administration in the United States tentatively approved the use of methandrostenolone (Dianabol) by clinicians. AAS research was, however, swiftly abandoned when Dr Ziegler discovered that his trial participants were taking as much as twenty times his instructed dosage regimen [Embleton & Thorne, 1998]. Widespread abuse of AAS by

American athletes later led to the Anabolic Steroid Control Act (1990) which named AAS as Schedule III controlled substances – essentially prohibiting their use within the United States.

Despite this prohibition, AAS use may now be considered ubiquitous among competition bodybuilders and many, including household names such as Arnold Schwarznegger, have admitted to abusing these drugs. Indeed, many competitions have arisen solely for drug-free bodybuilders such as the titles of Pro Natural World Champion, Pro Natural Mr. Universe and Pro Natural Mr. International. These competitions often impose a number of blood and urine tests as well as submitting competitors to a polygraph (lie detector) test. Despite this, many steroid-using bodybuilders try hard to evade a positive test result. One bodybuilder told me how, the day before a urine test, a friend of his had *"paid a tame doctor.. to push a tube [catheter] through his urethra.. to swap his urine [for that of someone without a history of AAS use]"*. Catheterization carries risks of urinary tract and kidney infections as well as bladder stones [Billote-Domingo *et al*]. The additional infection risks associated with exchanging urine in the manner described above are very easy to imagine. Clearly, with athletes going to such lengths to evade detection, it is very difficult for competition officials to ensure that competitors are drug-free even in natural bodybuilding competitions.

2.2.2
PREVALENCE OF AAS USE AMONG AMERICAN MEN

Although AAS first emerged among serious athletes, there is evidence that these products are now available to, and are being abused by, an increasing number of ordinary young men. Indeed, 6.3 percent of high school football players admit to having used AAS as do five percent of men entering New York gyms [Stilger *et al*, 1999]. Extrapolated to gyms across the country, this latter figure suggests that over five million American men are using or have recently used AAS [Kayayama *et al*, 2001]. Perhaps even more worrying is a report from The National Institute of Drug Abuse (NIDA) which estimates that around half a million 8th to 10th grade students have used AAS [Rannazzisi, 2004].

2.2.3
PRODUCTION AND DISTRIBUTION OF AAS

AAS production requires sophisticated pharmaceutical equipment and technical expertise. Most genuine AAS sold in the United States originate from countries where these drugs are available without prescription such as Germany, Thailand and Mexico. Most medical grade AAS sold illegally in the United States, however, are smuggled into the country from Eastern Europe. Other products are typically sold onto the black market by American physicians, pharmacists or veterinary surgeons.

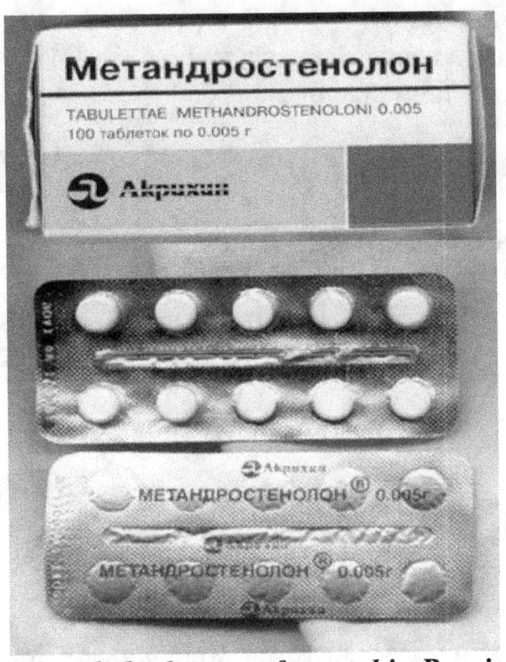

Dianabol tabs manufactured in Russia.

Table 1 represents the source of AAS used by respondents to this questionnaire. It shows that over half of AAS users obtain these products from international sources through online pharmaceutical websites. These include www.pharmanabolics.com/ and www.alinshop.in. An eight week supply of a cutting cycle from the latter source, for example, would cost a bodybuilder some $1,152 (£605). For the less committed, PharmAnabolics offers a 'beginners oral cycle' for only $450 (£236).

Table 1: Origin of AAS used by respondents	
Source	**Proportion of respondents**
Internet	56.5%
Friend	32.6%
Gym source	10.9%

Online drug sources also traditionally sell syringes – around $26 (£14) for 10 – which are additionally available from websites such as www.getpinz.com and www.gearboxinc.com. Many bodybuilders obtain syringes from addict needle exchange programs.

2.2.4
REASONS FOR USING AAS

One team of physicians has suggested that the question should not be 'why do men use AAS?' but 'why should men *not* use AAS'. Although side effects exist, they are often minimal and almost always reversible with further treatment. The abovementioned physicians liken AAS to a hypothetical drug for women which would a) not seriously damage their health, b) make them more beautiful within weeks, c) cause long-lasting results and d) was readily available by mail order. It is easy to see from this comparison why so many young men have used, or continue to use, AAS. In order to characterize what factors motivate men to use AAS, respondents were asked to assign a percentage to each of three reasons as to why they use AAS. These data are presented in Table 2.

Table 2: Reasons reported for using AAS		
Reason	**All respondents (%)**	**Respondents excluding competitive bodybuilders (%)**
Achieving status among peers	27	30
Attracting potential partners	30	32
Improving health or ability for a particular sport	48	46

As already noted, 34.4% (31/90) of those surveyed were already, or hoped to become, competition bodybuilders. As AAS use is an established part of competitive bodybuilding, the second column of Table 2 presents the responses of those men not pursuing bodybuilding as a career. Despite allowing for this distinction, the reasons given by 'ordinary' men did not differ significantly from those given by actual or hopeful competitive bodybuilders. Both sets of men assigned around 60 percent to cosmetic reasons for using AAS – namely achieving status and attracting potential partners.

2.2.5
HEALTH RISKS ASSOCIATED WITH AAS USE

As a result of the high cost and illegal nature of AAS, many sources have engaged in selling counterfeit products. The contents of these products may be harmless but there exists a significant risk that bodybuilders may inadvertently inject themselves with a toxic ingredient. One set of images, below, compares counterfeit and genuine Anavar tabs. One bodybuilder reported buying a bogus AAS product only to find later that the counterfeiter had not even spelled the drug name correctly on the label. Bodybuilders often communicate between each other to distinguish between authentic and counterfeit drugs. One man, for example, told how Dianobol tabs should break into large chunks when exposed to strong pressure between the thumbs. *"Fake tabs"*, he said, *"will just crumble"*.

Given the unregulated nature of AAS production, it is, of course, possible that even authentic drugs may contain toxic components. Many AAS products are synthesized in basement laboratories – often by athletes with little experience of the pharmaceutical industry. Some bodybuilders, for example, mentioned the Merck Index during unstructured interviews. This pharmacology manual provides instructions as to how many drugs, including AAS, may be synthesized in the laboratory [Embleton & Thorne, 1998]. A small shift in conditions such as temperature or pressure may, however, change the equilibrium of a reaction and result in an entirely different product to that expected. The potential dangers associated with this type of basement pharmacology do not need to be expounded here.

Many AAS are produced for veterinary purposes and so are not expected to meet the rigorous specifications and hygiene requirements of pharmaceuticals designed for human use. Indeed, almost 32 percent of respondents admitted to having used Equipoise; a veterinary steroid designed for horses. The image below shows a set of AAS products that are clearly labelled as being produced for veterinary use.

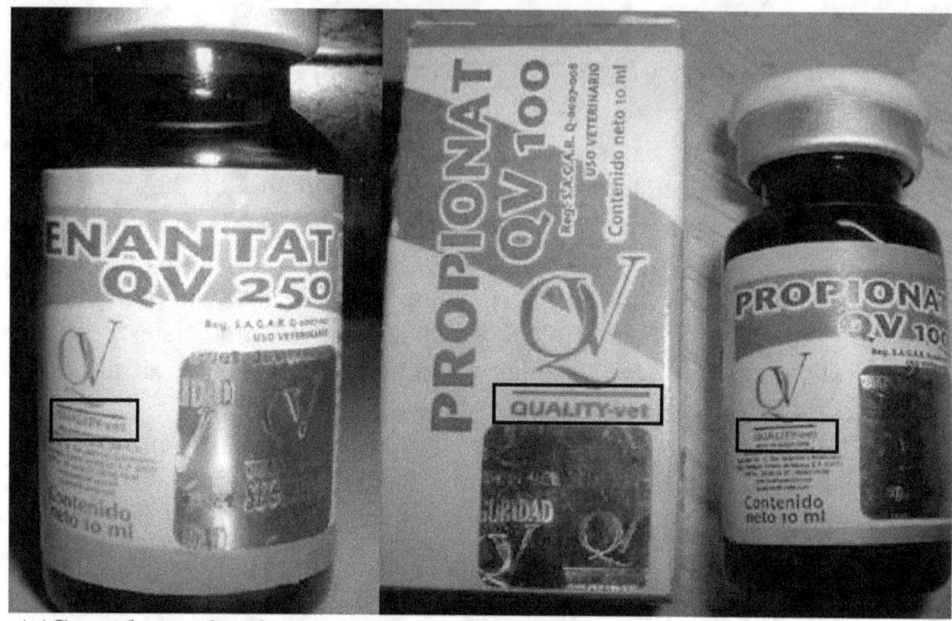

AAS products clearly marked for veterinary use

Even genuine AAS products may be associated with side effects. Some of these are represented in Table 3 together with the number of respondents reporting having experienced them. As a caveat it is, however, worth noting that AAS are prescribed by clinicians to prevent muscle wasting in both geriatric patients and in AIDS sufferers without any great risk to health [(Shroder *et al*, 2005) & (Grunfeld *et al*, 2006)]. Despite this, anecdotal evidence from bodybuilders and a number of studies have associated AAS with a number of side effects. Cosmetic side effects include a temporary reduction in testicular size, hair loss, acne and gynecomastia – the accumulation of 'female' breast tissue in the chest. The latter is caused by aromatization of testosterone in the body to oestrogen which is responsible for the development of many female secondary sexual characteristics. An example of non-steroid induced gynecomastia, which may be caused by puberty, a genetic condition or liver failure, is shown on the next page. Beyond these cosmetic side effects, there is also a strong possibility that steroid use may heighten relative risk of prostate cancer.

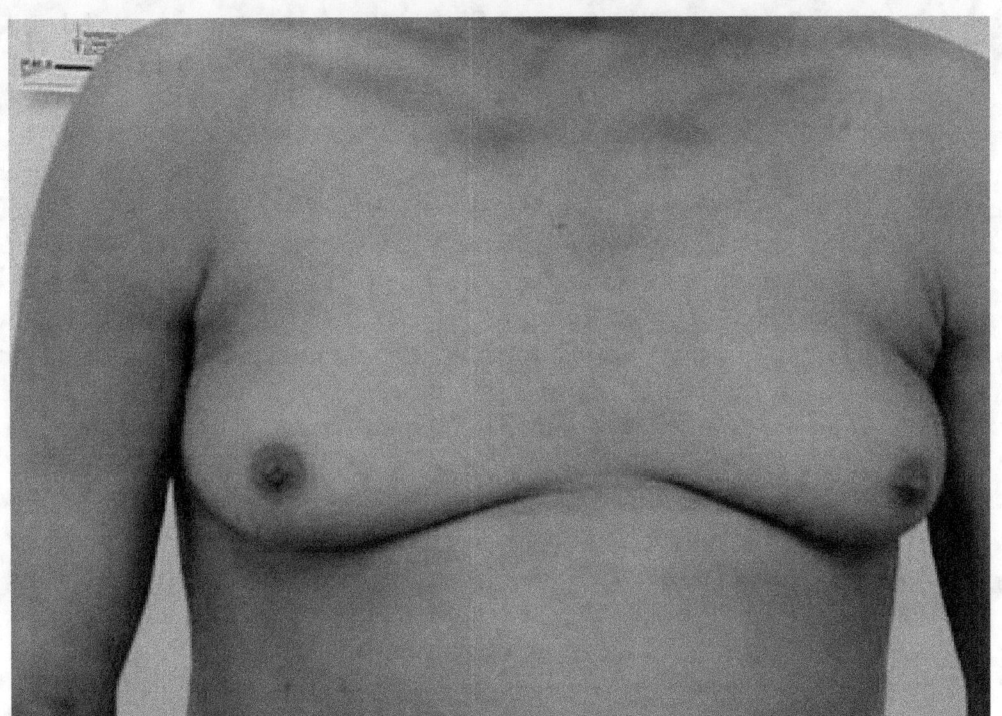

A severe case of gynecomastia (not induced by AAS use).
Courtesy and Copyright Dr Charles Goldberg, MD and the Regents of the University of California.

Table 3: Subjective side-effects reported by AAS users

Side effect	Incidence among AAS user respondents
Spontaneous aggression	27.5%
Sleeping disorders	25%
Steroid dependence	3.8%
Paranoia	2.5%
Depression	11.2%
Gynechomastia	9%
Testicular atrophy	12.5%
Acne	68.8%
Hair loss	12.5%

Psychiatric disturbances have also been attributed to AAS. These include confusion, hallucinations, onset of sleeping disorders, depression on withdrawal, spontaneous aggression ("roid rage") and psychosis. Of these, the aggressive effects of AAS have proven most controversial and have been hotly disputed by some researchers [(Fudala *et al*, 2003) & (Tricker *et al*, 1996)]. Despite this, almost 30 percent of survey respondents reported having experienced spontaneous aggression or temper tantrums which they attributed to use of AAS. One interviewee described *"roid rage"* as *"that feeling you get in the shopping mall where you're just waiting for someone to push you so you can smash their head in"*. Many others described experiences of spontaneous aggression. Robert, a twenty year-old from Los Angeles, overfilled his blender when making a protein shake so that the blades could not spin properly. *"Then I lost it"*, he said, *"and smashed the blender on the floor.. immediately regretting my futile action, I began cleaning up the mess.. cut myself on the glass.. lost it again and laid into the fridge"*. Another man who described himself as *"not a violent guy"* told how he lost his temper with two men in a club, *"blacked out and came around to find them both out cold; one with a broken leg and the other with his face smashed in"*. This man did not recall inflicting these injuries but witnesses watched him do so nonetheless. One focus group participant, however, disputed this anecdotal evidence. He explained that irritation and aggression are known to afflict all athletes who over-train and these instances of *"roid rage"* may be better explained as a symptom of AAS users pushing themselves too hard with their weight training. Although some researchers agree with this interpretation, their research is limited as, ethically, they are only permitted to administer up to 300mg testosterone per week to their study participants. Respondents to this survey were, however, were using, on average, over 1000mg in this same timescale. Other studies using AAS users, instead of laboratory participants, have found that over 23 percent report major psychiatric disturbances [Pope & Katz, 1994]. This figure is similar to the values reported in Table 2 which suggests that these symptoms are genuinely experienced by men self-administering high doses of testosterone.

2.2.6
AAS AND OTHER BODY IMAGE DRUGS

AAS may also prove dangerous because of their association with a cocktail of other drugs. Bodybuilders frequently use combinations (so-called 'stacks') of AAS products during any one cycle. One eight week cutting cycle for example included 56 x 50mg ampoules of Winstrol, 5 x 10ml vials of Trenabol, 30ml testosterone propionate, 2 x 10ml vials of Mastabol, 200 x 20mcg tabs of Clenbuterol, 120 x 50mcg tabs of Cytomel, 100 x 0.25mg tabs of Arimidex, 2 x 5000iu HCG, 50 x 10mg tabs of Zenneca and 30 tabs of Clomid. Clearly, if AAS are potentially dangerous, then consumption of this number and in these doses over only eight weeks is bound to amplify the health risk. Many of these products are, however, not steroids and instead represent an attempt on behalf of the user to minimize side effects. Table 4 summarizes the drugs taken by respondents as a part of what is known to steroid users as post-cycle therapy (PCT). These drugs may however be associated with their own side effects particularly as their use is unregulated and not under the guidance of a competent physician.

Table 4: Drugs used by respondents during post-cycle therapy (PCT)

Product	Side effects reduced	Potential new side effects	Incidence of use
hCG	Reduced endogenous testosterone	Minimal - acne, blood clots	24%
Anti-oestrogens	Gynecomastia	Stroke, blood clots	74%
Anti-depressants	Depression	Dependence	8%
hGH	Muscle loss	Hypoglycemia, reduced thyroid function	10%
Insulin or IGF-1	Diabetic symptoms	Hypoglycemia, brain damage, coma, death	7%
Clenbuterol	Muscle loss	Heart degeneration	8%

Human chorionic gonadotrophin (hCG), the female hormone detected by pregnancy tests, was reportedly used by 24 percent of respondents to stimulate endogenous testosterone production. Without this stimulus, the body may come to rely on the injected testosterone and so reduce its natural production of this hormone. Although hCG may aggravate acne and increase the risk of blood clots, it is not known to cause other serious side effects.

Another class of side effect-reducing drugs are the anti-oestrogens. These include the breast cancer drug tamoxifen which blocks oestrogen receptors to prevent side effects such as gynecomastia. This is important for bodybuilders as gynecomastia interrupts their pursuit of a hypermasculine physique but also because it is a clear suggestion to other weight lifters of steroid abuse. As a result, over 76 percent of steroid users surveyed admitted to having used anti-oestrogens. Tamoxifen is, however, known to increase risk of blood clots and stroke [Mouriden, 2006]. There is, however, an alternative to using anti-oestrogens. One 28 year-old bodybuilder had gone as far as to undergo surgery to remove his *"gyno"* which he recommended as *"painless.. and well worth it"*.

Almost eight percent of respondents reported having used anti-depressants in order to combat withdrawal symptoms following an AAS cycle. Insulin was used by almost seven percent of respondents for its muscle building properties and to prevent the diabetic effects sometimes caused by AAS use. Although inappropriate use of insulin can cause hypoglycemia, coma, irreversible brain damage and death, respondents seemed more concerned that insulin also promotes fat storage. For this reason respondents counselled use of a blood glucose monitor and to *"avoid all fatty foods.. during the active life of the slin [insulin]"*. Insulin is, of course, readily available on the black market because of its distribution to diabetics.

Human growth hormone (hGH) was used by over ten percent of respondents for its anabolic properties. As a body image drug, hGH is little short of a miracle as it not only promotes muscle growth but also reduces signs of aging. One author has claimed that bodybuilders can gain as much as forty pounds of muscle in only ten weeks by injecting sufficiently large quantities of hGH [Roberts, 2004]. Without drugs, a committed bodybuilder could normally only hope to achieve 2lbs of lean muscle per month. Despite this achievement, hGH is undoubtedly the most expensive bodybuilding supplement on the black market at almost $170 per day of use.

Although not an AAS, clenbuterol can also promote muscle growth. Of all the supplements discussed with bodybuilders in LA, clenbuterol caused the most fear as it is perceived by the community as being particularly dangerous. Many men advised that only competition bodybuilders should consider using clenbuterol during a steroid stack. Despite this advice, almost eight percent of questionnaire respondents admitted to having used clenbuterol. The dangers of this drug described by interviewees are strongly supported by the medical literature. Rats treated with clenbuterol exhibit greatly enlarged hearts, largely due to collagen infiltration, and cardiac cell degeneration. Collagen infiltration reduces the flexibility of cardiac tissue and is likely to reduce cardiac output [Burniston *et al*, 2002]. Clenbuterol has also been shown to reduce aerobic performance in horses after only eight weeks of use [Kearns & McKeever, 2002]. For these reasons clenbuterol is considered a very dangerous drug with side effects which far outweigh its muscle building properties.

2.2.7

AAS AND RECREATIONAL DRUGS

A number of studies have further associated AAS with the use of amphetamines, cannabis, cocaine and injectable drugs [(DuRant *et al*, 1993) & (Korkia, 1996)]. The suggestion here is that users are driven to obtain AAS from drug dealers who may also provide additional recreational drugs. It is also possible that learning to self-administer injectable AAS may increase the risk of later using drugs such as heroin. This phenomenon has been dubbed becoming "needle happy". Responses to this survey, however, suggest a very different relationship between AAS use and abuse of recreational drugs. Table 5 illustrates this observation by showing the proportion of respondents admitting to drug use before and after their first cycle of AAS.

Table 5: Use of recreational drugs by AAS users		
Drug	**Incidence of use before first AAS cycle**	**Incidence of use after first AAS cycle**
Amyl nitrate	2.6%	1%
Cocaine	8.9%	3.6%
Crack cocaine	2.6%	1%
Crystal meth	2.6%	1%
Ecstasy	10.7%	2.7%
Heroin	1%	1.8%
LSD	7.1%	1.8%
Marijuana	30.4%	9.8%
Speed	6.3%	1%

Table 5 and the following bar chart clearly show that, with the exception of heroin, reported drug use dramatically declined once respondents had begun to use AAS. One respondent explained this as a symptom of his newly adopted 'healthy' lifestyle which he more strongly identified with once taking AAS to improve strength and body image. Indeed, little more than seven percent of respondents reported that their source of AAS also sells recreational drugs and, furthermore, over 96% denied having bought recreational drugs from their primary AAS supplier. These data cast some doubt on previous findings that AAS are associated with increased use of recreational drugs.

The majority of respondents (64%) reported self-administering their first steroid cycle by means of intramuscular injection. This proportion increased to 71 percent (65/91) for their most recent steroid cycle. Injectable drugs are considered to be more efficient (because of higher bioavailability) and safer by users, as oral products are extensively metabolized by, and so may cause damage to, the liver. Users typically push the needle deep through the skin and draw back the plunger until blood cannot be seen in the syringe. This is accepted as confirmation that the needle has reached the muscle.

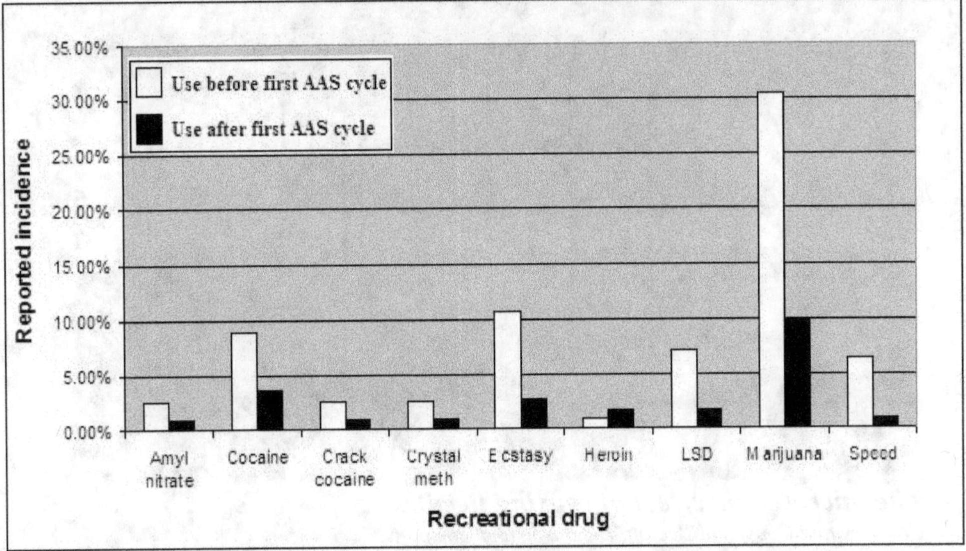

The use of recreational drugs before and after the first AAS cycle

Although intramuscular injection is thought to be safer than intravenous administration, injectable AAS may increase risk of minor infections and more serious diseases such as HIV/AIDS and chronic viral hepatitis. Indeed, previous researchers have found that as many as 13 percent of AAS users use unsafe injection techniques such as reusing needles, sharing needles or using multi-dose vials [Parkinson & Evans, 2006]. During the course of this project, two bodybuilders were overheard discussing the most efficient technique of 'sterilising' needles. Although these men were not sharing needles with other people, re-use damages the needle and significantly depletes its lubrication. The result of needle reuse is illustrated by the following image. Reused needles cause more injection pain, bruising and are prone to snapping off underneath the skin.

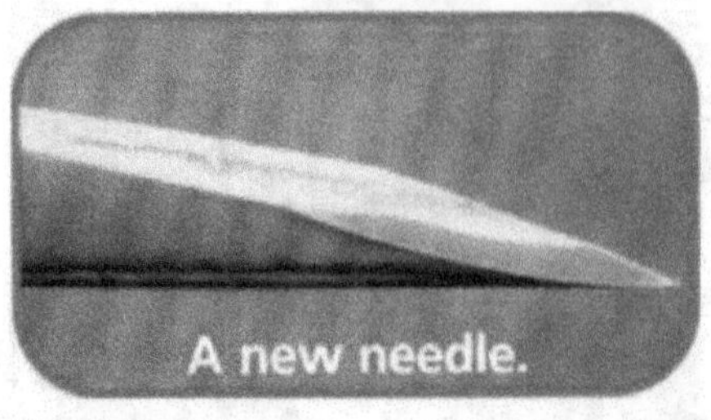

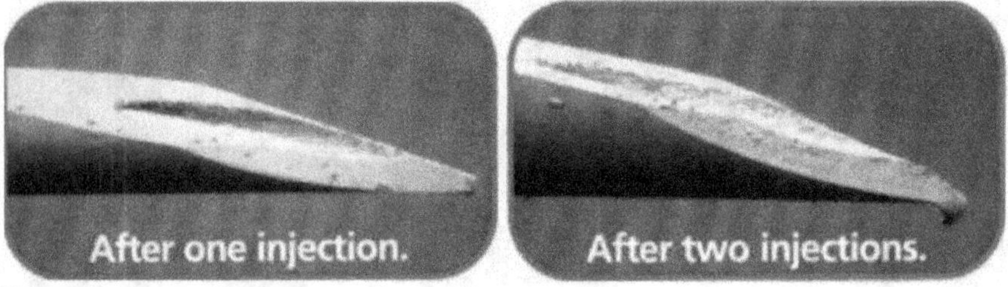

The microscopic effect of reusing needles.
Courtesy and Copyright Becton, Dickinson and Company.

Aside from reuse, improper delivery of an intramuscular injection is also likely to expose users to the risk of localized infection which in some cases may result in surgery to drain an abscess [Wynaden *et al*, 2005]. Intramuscular injections into the buttock have additionally been known to damage the sciatic nerve, resulting in lower limb palsy [Mayer & Romain, 2001]. Clearly, then, it is not ideal for untrained men to be self-administering intramuscular injections without the supervision of a healthcare professional.

2.2.8
CONCLUSION

Although the potential health consequences of AAS use are numerous, they are not yet well-defined enough to discourage young men from abusing these drugs. Indeed, their perceived benefits are such that young men are becoming increasingly likely to seek suppliers of AAS products. This is particularly concerning as AAS use is often associated with abuse of other, potentially more dangerous, body image drugs. Despite the risks, it seems likely that AAS use will continue so long as young men are presented with impossibly lean and muscular 'ideal' physiques by an appearance-obsessed media.

2.3
Psychiatric Body Image Disorders

Closely following the general increase in male body dissatisfaction have come the psychiatric pathologies associated with body image distortion. The better known of these illnesses are the eating disorders anorexia nervosa and bulimia nervosa. Less well known, but increasing in prevalence no less quickly, is body dysmorphic disorder (BDD).

2.3.1
BODY DYSMORPHIC DISORDER

As previously described, body dysmorphic disorder (BDD) is a distressing or impairing preoccupation with a nonexistent or slight defect in body appearance [Phillips *et al*, 1994]. Many sufferers of this disorder become housebound, self-harm or attempt suicide. Of sixteen patients presenting to a dermatologist and who later went on to commit suicide, a disproportionate number suffered from BDD [Cotterill & Cunliffe, 1997]. As one man wrote, when he is afflicted by an 'attack', *"I start to get sad and cry and dream of suicide and how I am really ugly deep down inside... I would like to just end my life at these times"*. Another described his situation thus: *"I want this shit to stop... I wish I had killed myself in December like I planned... but like everything else in my life - I can't complete anything"*.

BDD is often accompanied by obsessive behaviours such as questioning, grooming and mirror checking. One paper describes a bodybuilder who asked his father twenty times in one hour whether or not he was looking any more muscular [Pope *et al*, 1997]. Another focus group participant reported that *"[he] wouldn't be surprised if the number of times per day [he] look[s] in it [the mirror] was in the triple digits"*. Another wrote that *"the mirror is my enemy... I can't look at myself without immediately examining all of the things I hate about my looks"*. And another, *"I hate mirrors with a vengeance and yet I am compelled to stare into them"*.

BDD affects as many men as it does women [Phillips & Diaz, 1997]. Despite this, the sexes are affected differently with male sufferers more likely to exhibit co-morbid bipolar disorder than their female counterparts. Men are also more likely to be preoccupied with their genitals, height, excessive body hair and muscularity whereas women typically emphasise breast and leg concerns [Perugi *et al*, 1997]. One man, for example, described an obsession with a lump on his forehead which he was convinced other people would think was a tumour. *"I have compulsions where I stare in reflective surfaces..."*, he wrote, *"...whenever I watch TV I get fixed on other people's foreheads. I'm even obsessed with an actor's forehead, Jean Claude Van Damme, who appears to have a lump on his forehead... I watch his movies and stare at his forehead. I can't believe how he continues to not care about [it]"*. This man further suffered from generalised anxiety, panic attacks and had misled his surgeon in order to have his 'lump' removed.

Following the recognition of BDD as a pathological disorder, a team of psychiatrists began to identify cases of muscle dysmorphia. In this variant of BDD, sufferers become pathologically preoccupied with their muscularity and adopt an *"all-consuming lifestyle revolving around their work-out schedule and meticulous diet"* [Pope *et al*, 1997]. Comparison of muscle dysmorphic men with healthy bodybuilders suggests that the former differ greatly from the latter – most of whom do not exhibit psychopathology [Olivardia *et al*, 2000]. Muscle dysmorphia is very often also associated with other BDD symptoms. As one body dysmorphic man wrote, *"[my area of concern is] all my Face uurrrgghhhh, Biceps, Deltoids, Chest... Arse"*. Nevertheless, men suffering from muscle dysmorphia are more likely to attempt suicide and to abuse illicit substances such as AAS than are other BDD sufferers [Pope *et al*, 2005].

BDD in men may result in hugely diminished self-esteem. *"All I know is I hate myself..."*, wrote one man, *"...I feel I am...genetically inferior. You can't imagine how bad I look sometimes... I'm the runt of my family"*. Another man, a junior doctor in surgical training, described how he *"looks[s] in the mirror and [is] thoroughly thoroughly disgusted by what [he] see[s]"*. While he usually *"pretends to be happy and positive all the time"*, he wrote that, in reality *"I cry myself to sleep and wish a bus would hit me and put an end to this suffering"*. BDD by definition results in huge social impairment. As a twenty year-old American man reported, *"I make sure I have no friends too and isolate myself because I think people deserve better...they shouldn't condescend to such an ugly loser like me... and I don't want to make them look bad"*.

Effective treatments for BDD appear to be limited to pharmacological interventions using high dosage serotonin reuptake inhibitors [Phillips, 2000]. Some forms of cognitive therapy have also been shown to improve the mental health outcomes of patients suffering from BDD [(Phillips & Castle, 2001) & (Penzel, 2000)]. Many sufferers first present to a dermatologist or cosmetic surgeon before even considering that they might be suffering from a psychiatric disorder. One young Belgian homosexual man for example reported that he had already undergone a *"hair transplant, had a few fatgrafts, chin implant, nose jobs, etc"* by the age of twenty. Recent evidence, however, suggests that attempted surgical or dermatological resolution of the presenting complaint typically fails to lessen the underlying BDD [Crerand *et al*, 2005]. The Belgian man previously described wrote of his surgery that *"it has helped me a bit but it's definitely not THE answer but I'm still convinced that it can make a difference"*.

2.3.2
ANOREXIA NERVOSA AND BULIMIA NERVOSA

Anorexia nervosa is an eating disorder characterised by a pathological loss of appetite usually following a severe body image disturbance [American Psychiatric Association, 2000]. Two decades ago the ratio of anorexia in males and females was thought to be 1:15 [Ziora *et al*, 2006]. This has now increased so that men accounted for as many as ten percent of anorexia nervosa diagnoses in 1990 [Barry & Lippman, 1990] and up to 25 percent of cases diagnosed today [Ziora *et al*, 2006]. The disorder is usually accompanied by inappropriate fasting and excessive cardiovascular exercise. One man, for example, reported running six miles every time he ate something he considered to be fattening. Another had driven his weight as low as 42kg.

Bulimia nervosa is an eating disorder characterised by cycles of binge eating followed by inappropriate purging to avoid weight gain [American Psychiatric Association, 2000]. These may include self-induced vomiting, fasting or use of laxatives or diuretics. Bulimia is also on the rise among the male population. One American bulimic man reported that, on residential treatment programmes he has attended, most of his fellow sufferers were involved in sports such as wrestling *"where weight is always an issue"*. This raises the possibility of a further association with bodybuilding. Although competitive bodybuilders share many of the behaviours associated with bulimics, they have been found to share few of the psychological difficulties characteristic of this latter group [Goldfield *et al*, 2006].

2.3.3
OBSESSIVE-COMPULSIVE DISORDER

Obsessive compulsive disorder (OCD) is typically characterised by obsessive, intrusive thoughts which are related to compulsions in order to neutralise these obsessions [American Psychiatric Association, 2000]. In terms of its relationship to muscle dysmorphia, a sufferer of this latter disorder is typically preoccupied with the idea that he is insufficiently muscular. In order to neutralise these obsessions, he compulsively lifts weights, pays excessive attention to diet and may become socially impaired. In this way, muscle dysmorphia may be more closely related to OCD than it is to BDD [Chung, 2001].

2.3.4
CONCLUSION

One debate in the field of body image research centres around the classification of muscle dysmorphia. Many have, for example, suggested that muscle dysmorphia should be considered to be a predominantly male equivalent of anorexia nervosa. This has led to the colloquial naming of muscle dysmorphia as 'reverse anorexia' and 'bigorexia' [Pope *et al*, 2002]. Indeed, bodybuilders exhibit an anorexia lifetime prevalence of around 8.3% compared to only 0.02% in the general population [Pope *et al*, 1993]. Others have, however, preferred to see muscle dysmorphia as a variant of BDD. Evidence for this comes from analysis of bodybuilders in South Africa – 33% of which suffered from BDD-induced appearance obsessions other than muscularity [Hitzeroth *et al*, 2001]. More recently, it has been proposed that the clinical presentation of muscle dysmorphia is more consistent with its classification as an obsessive-compulsive disorder [Chung, 2001].

Unlike the Adonis Complex in which men aspire to impossible degrees of attractiveness, eating disorders in men typically prey on distorted body perception rather than on a desire to reach impossible ideals [Mangweth *et al*, 2003]. This has been demonstrated by asking man with eating disorders to select images which best represent their current appearance and those which represent their ideal appearance. Men with eating disorders chose images twice as obese as themselves to represent their current physique. This pattern was not observed in a similar sample of male mountain climbers or a sample of male controls.

Despite this difference between the Adonis Complex and pathological eating disorders, the two remain inseparable. It is difficult to disregard, for example, the observation that male body dissatisfaction among the general population has increased alongside eating disorders. Indeed, questionnaires administered to college students have shown that general body dissatisfaction is closely associated with depression, low self-esteem and eating pathology [Olivardia *et al*, 2004].

2.4

Summary

1. Male appearance concerns have fuelled an industry which results in adherence to obsessive rituals of weight training, eating and sleeping. Events which prevent men from meeting their training goals may result in feelings of guilt or anger.

2. Many men further use over-the-counter body image drugs. The health consequences of using large doses of these substances for long periods of time are as yet unknown.

3. Others go as far as to abuse illicit substances such as anabolic androgenic steroids. These drugs have been associated with a number of physiological and psychiatric disorders. Steroid users often use a cocktail of other dangerous drugs such as insulin and clenbuterol in order to build as much lean muscle as possible.

4. A growing proportion of men are suffering from eating disorders such as anorexia nervosa and bulimia nervosa. This is likely to be related to the general decline in body dissatisfaction among men described in Part I.

5. Men may alternatively develop body dysmorphic disorder in which they become pathologically preoccupied with the perceived ugliness of a particular element of their body. When this element is their perceived lack of muscularity, the disorder is known as muscle dysmorphia.

6. BDD and/or muscle dysmorphia are risk factors for social exclusion, unemployment, depression, self-harm and suicide.

Afterward

I am listening to a politician speak. He is telling his assembled audience about his ascent to greatness: of his success in Hollywood and, more recently, in government. All of this, from his starring roles in thirty movies to his landslide victory in the race to the Governor's mansion, he attributes to the day when, at fifteen years of age, he began weight training. *"It gave me the foundation, the contacts and the confidence"*, said Mr Schwarzenegger, *"for everything I would later go on to achieve"*.

Throughout this monograph I have emphasised the tragic consequences of men trying to improve their appearance. These have included social impairment, illegal drug use, and psychiatric illness leading to self harm and suicide. These tragedies are not, however, experienced by the vast majority of men who invest effort in improving their appearance. Indeed, men have always worried to some degree that their appearance is consistent with the latest cultural expectations. In the seventeenth century, when portliness was considered the ideal male physique, some men took to stuffing their shirts with bran and rags [Fairholt, 1860]. This is surely analogous, in many ways, to the weight training efforts of some contemporary men. As one group of researchers warn, *"muscle dysmorphia should not be confused with mere enthusiasm for bodybuilding... ordinary dedication to sports is not associated with the profound body dissatisfaction, subjective distresses and impaired social and occupational functioning reported by individuals with.. muscle dysmorphia"* [Pope *et al*, 1991].

Mr Schwarzenegger addresses an audience of bodybuilders.

Of the many men contacted during this project, none even so far as suggested that bodybuilding had had an overall negative impact on their lives. Even those who might have fallen into a diagnosis of muscle dysmorphia saw weight training as a therapy which helped suppress their feelings of inadequacy. In terms of bodybuilding itself, men variously felt that it had improved their confidence, self-esteem, success with the opposite sex, ability to relieve stress and their popularity. Many led healthier lives – consuming less alcohol, sugar and saturated fats – as a direct result of their efforts invested in the gym. Indeed, it is for these reasons that Mr Schwarzenegger cycles past Muscle Beach each week with his children. *"I want them to learn the importance"*, he told his audience, *"of always striving for improvement... I learned that from bodybuilding"*.

It is, however, clear that men are becoming increasingly dissatisfied with their appearance. In itself, this is not necessarily a problem. Indeed, they may simply be catching up with women in this regard who have met with aesthetic cultural expectations for centuries. The real problem arises when, like women, men are exposed to an 'ideal' physique which cannot be achieved without recourse to illegal and potentially harmful drugs. Over and above this, however, men may be at even greater risk that women. Men are, for example, less likely to seek help when they first perceive that their appearance concerns might be psychologically harmful. Although men are more reluctant to visit health professionals than women [Chapel *et al*, 2004], this is particularly true for men suffering from pathological appearance concerns [Pope *et al*, 2002]. Furthermore, when men *do* present to clinicians with these concerns, they are often not taken seriously. When one BDD sufferer first mentioned his problem to a therapist, he perceived that they were laughing at him. As a result, he did not discuss his BDD symptoms with another professional until many years later.

Clearly these problems must be addressed. If our culture is to become more appearance-orientated, then notions of masculinity must shift to accept that change. Furthermore, the signs and symptoms of BDD should be circulated more widely among clinicians so that this illness can be diagnosed and treated when men present to a dermatologist, cosmetic surgeon or general practitioner. Finally, our print and film media must be pressured into limiting their use of unobtainable physiques, whether these be anorexic women or steroid fuelled men. In this way, both men and women might hope to regain a sense of proportion when considering the value of their own appearance.

- 84 -

References

INTRODUCTION

American Psychiatric Association. (2000). *Diagnostic and Statistical Manual of Mental Disorders*, 4th ed.

Bohne, A., Keuthen, N. J.,Wilhelm, S., Deckersbach, T. & Jenike, M. A. (2002). Prevalence of symptoms of body dysmorphic disorder and its correlates: a crosscultural comparison. *Psychosomatics* 43;486-490.

Burkert, W. (1985). *Greek Religion.* Published by Harvard University Press, p176.

Clausen, S. (2005). Body dysmorphic disorder: through a glass darkly. *Student British Medical Journal* 13;133-176.

Corson, P. M. & Andersen, A. E. (2002). Body image issues among boys and men in Cash, T. F. & Pruzinsky, T. (eds.) *Body image: a handbook of theory, research and clinical practice.* Published by The Guildford Press.

Embleton, P. & Thorne, G. (1998). *Anabolic Primer.* Published by MuscleMag International.

Olivardia, R., Pope, H. G. & Hudson, J. L. (2000). Muscle dysmorphia in male weightlifters: a case-control study. *American Journal of Psychiatry* 157;1291-1296.

Phillips, K. A. (1996). *The broken mirror: understanding and treating body dysmorphic disorder.* Published by Oxford University Press. p131.

Phillips, K. A. & Castle, D. J. (2001). Body dysmorphic disorder in men: psychiatric treatments are usually effective. *British Medical Journal* 323;1015-1016.

Pope, H. G., Phillips, K. A. & Olivardia, R. (2002). *The Adonis Complex: how to identify, treat and prevent body obsession in men and boys.* Published by Simon & Schuster Inc.

BODY IMAGE CONCERNS IN MEN

Agliata, D. & Tantleff-Dunn, S. (2004). The impact of media exposure on males' body image. *Journal of Social and Clinical Psychology* 23(1);7-22.

Alfonso, M., Richter-Appelt, H., Tosti, A., Sanchez Viera, M. & Garcia, M. (2005). The psychosocial impact of hair loss among men: a multinational European study. *Current Medical Research and Opinion* 21(11);1829-1836.

Berscheid, E., Walster, E. & Bohrnstedt, G. (1973). The happy American body: A survey report. *Psychology Today* 7;119–131.

Cash, T. F., Gillen, B. & Burns, D. S. (1977) Sexism and 'beautyism' in personnel consultant decision making. *Journal of Applied Psychology* 62; 301-310.

Cash, T. F., Winstead, B. W. & Janda, L. H. (1986). The great American shape-up: Body image survey report. *Psychology Today*, 20(4), 30-37.

Clark, M. S. & Mills, J. (1979). Interpersonal attraction in exchange and communal relationships. *Journal of Personality and Social Psychology* 37;12-24.

Francken, A. B., van de Wiel, H. B. M., van Driel, M. F. & Weijmar Schultz, W. C. M. (2002). What importance do women attribute to the size of the penis? *European Urology* 42;426-431.

Goleman, D. (1991). When ugliness is only in the patient's eye, body image can reflect mental disorder. *New York Times*, 2nd October.

Gupta, M. A. & Gupta, A. K. (1998). Depression and suicidal ideation in dermatology patients with acne, alopecia areata, atopic dermatitis and psoriasis. *British Journal of Dermatology* 139(5);846-850.

Harper, B. (2000). Beauty, statute and the labour market: a British cohort study. *Oxford Bulletin of Economics and Statistics* 62;773-802

Lasek, R. & Chen, M. (1998). Acne vulgaris and the quality of life of adult dermatology patients. *Archives of Dermatology* 134(4);454-458.

Lee, P. A. & Reiter, E. O. (2002). Genital size: a common adolescent male concern. *Adolescent Medicine* 13(1);171-180.

Lewis, R. W. & Witherington, R. (2005). External vacuum therapy for erectile dysfunction: use and results. *World Journal of Urology* 15(1);78-82.

Mondaini, N., Ponchietti, R., Gontero, P., Muir, G. H., Natali, A., Di Loro, F., Caldarera, E., Biscioni, S. & Rizzo, M. (2002). Penile length is normal in most men seeking penile lengthening procedures. *International Journal of Impotence Research* 14(4);283-286.

Olivardia, R., Pope, H. G. & Hudson, J. L. (2000). Muscle dysmorphia in male weightlifters: a case-control study. *American Journal of Psychiatry* 157;1291-1296.

Park, M. J., Kang, Y. J. & Kim, D. H. (2003). Dissatisfaction with height and weight, and attempts at height gain and weight control in Korean school-children. *Journal of Pediatric Endocrinology and Metabolism* 16(4);545-554.

Passchier, J., Erdman, J., Hammiche, F. & Erdman, R. A. (2006). Androgenetic alopecia: stress of discovery. *Psychological Reports* 98(1);226-228.

Phillips, K. A. (1996). *The Broken Mirror: Understanding and treating body dysmorphic disorder*. Published by Oxford University Press. p193-208

Phillips, K. A. & Diaz, S. F. (1997). Gender differences in body dysmorphic disorder. *The Journal of Nervous and Mental Disease* 185(9);570-577

Pope, H. G., Phillips, K. A. & Olivardia, R. (2002). *The Adonis Complex: how to identify, treat and prevent body obsession in men and boys*. Published by Simon & Schuster Inc.

Raevuori, A., Keski-Rahkonen, A., Bulik, C. M., Rose, R. J., Rissanen, A. & Kaprio, J. (2006). Muscle dissatisfaction in young adult men. *Clinical Practice and Epidemiology in Mental Health* 2;6-14.

Small, M. F. (1996). *What's love got to do with it? The evolution of human mating.* Published by Anchor Publishing Ltd.

Wessells, H., Lue, T. F. & McAninch, J. W. (1996). Penile length in the flaccid and erect states: guidelines for penile augmentation. *Journal of Urology* 156;995-997.

Van Driel, M. F., Weijmar Schultz, W. C. M., Van de Wiel, H. B. M. & Mensink, H. J. A. (1998). Surgical lengthening of the penis. *British Journal of Urology* 82;81-85.

PSYCHOSOCIAL CAUSES OF THE ADONIS COMPLEX

Andersen, A. E. & Didomenico, L. (1992). Diet vs. shape content of popular male and female magazines: a dose-response relationship to the incidence of eating disorders. *International Journal of Eating Disorders* 11(3);283-287.

British Medical Association. (2000). *Eating disorders, body image and the media.* BMJ Books.

BodyBuildingPro.com. Biography and Contest History of Arnold Schwarzenegger.
http://www.bodybuildingpro.com/arnoldschwarzenegger.html

Brownell, K. D. & Napolitano, M. A. (1995). Distorting reality for children: body size proportions of Barbie and Ken dolls. *International Journal of Eating Disorders* 18(3);295-298.

Gillet, J. & White, P. G. (1992). Male bodybuilding and the reassertion of hegemonic masculinity: a critical feminist perspective. *Play Culture* 5;358-369

Hartlaub, P. (2003). Before he was a candidate Schwarzenegger openly flexed his vocal cords. *San Francisco Chronicle*. Friday, August 15, 2003.

Harrison, K. & Cantor, J. (1997). The relationship between media consumption and eating disorders. *Journal of Communication* 47(1);40-67.

Howe, A., Owen-Smith, V. & Richardson, J. (2003). Media influence on suicide. *British Medical Journal* 326;498.

Jade, D. (2002). *Eating disorders and the media*. National Centre For Eating Disorders.

Leit, R. A., Pope, H. G. & Gray, J. J. (1999). Cultural expectations of muscularity in men: the evolution of playgirl centrefolds. *International Journal of Eating Disorders* 29(1);90-93.

Mondaini, N., Ponchietti, R., Gontero, P., Muir, G. H., Natali, A., Di Loro, F., Caldarera, E., Biscioni, S. & Rizzo, M. (2002). Penile length is normal in most men seeking penile lengthening procedures. *International Journal of Impotence Research* 14(4);283-286.

Neimark, J. (1994). The beefcaking of America – sexual objectification of men by women. *Psychology Today,* Nov/Dec '94

Pope, H. G., Olivardia, R., Gruber, A. & Borowiecki, J. (1999). Evolving ideals of male body image as seen through action toys. *International Journal of Eating Disorders* 26(1);65-72.

Pope, H. G., Olivardia, R., Borowiecki, J. J. & Cohane, G. H. (2001). The growing commercial value of the male body: a longitudinal survey of advertising in women's magazines. *Psychotherapy and Psychosomatics* 70;189-192.

Pope, H. G., Phillips, K. A. & Olivardia, R. (2002). *The Adonis Complex: how to identify, treat and prevent body obsession in men and boys.* Published by Simon & Schuster Inc.

Schwarzenegger, A. (1977). *Arnold: Developing a Mr. Universe Physique*. Mail-order pamphlet.

Schwarzenegger, A. (2003). *Arnold: The education of a bodybuilder*. Published by Simon & Schuster Inc.

Terzief, J. (2006). Fashion world says too thin is too hazardous. *Womens e-News*, 24th October.
http://www.womensenews.org/article.cfm/dyn/aid/2899/context/cover/

Wolf, N. (1992). *The Beauty Myth: how images of beauty are used against women*. Published by Anchor Inc.

Yang, C. F., Gray, P. & Pope, H. G. (2005). Male body image in Taiwan versus the West: Yanggang Zhiqi meets the Adonis Complex. *American Journal of Psychiatry* 162;263-269.

SOCIOBIOLOGICAL CAUSES OF THE ADONIS COMPLEX

Campbell, B. C., Pope, H. G. & Filiault, S. (2005). Body image among Ariaal men from Northern Kenya. *Journal of Cross-Cultural Psychology* 36(3);371-379.

Diamond, J. (1992). *The rise and fall of the third chimpanzee*. Published by Vintage Inc. pp59-69.

Dixson, A., Dixson, B. & Anderson, M. (2005). Sexual selection and the evolution of visually conspicuous sexually dimorphic traits in male monkeys, apes and human beings. *Annual Review of Sex Research* 16;1-19.

Geissman, T. (1998). Body weight in wild gibbons (Hylobatidae). In: *Abstracts, XVIIth Congress of the International Primatological Society*, Aug 10-14. Abstract 282.

Kratochvil, L. & Frynta, D. (2002). Body size, male combat and the evolution of sexual dimorphism in eublepharid geckos. *Biological Journal of the Linnean Society* 76(2);303-314.

Lipinski, J. P. & Pope, H. G. (2002). Body ideals in young Samoan men: a comparison with men in North America and Europe. *International Journal of Men's Health* 1(2);163-171.

Long, K. C. (2006). Primate Fact Sheets: Gorilla gorilla sp. *Primate Info Net Library and Information Service.* http://pin.primate.wisc.edu/factsheets/entry/gorilla

McCann, T. S., Fedak, M. A. & Harwood, J. (1989). Parental investment in southern elephant seals, Mirounga leonine. *Journal of Behavioral Ecology and Sociobiology* 25(2);81-87.

Ogden, C. L., Fryar, C. D., Carroll, M. D. & Flegal, K. M. (2004). Mean body weight, height and body mass index, United States 1960-2002. *Advance Data from Vital and Health Statistics*, Centre for Diseases Control 347.

Wilson, E. O. (2000). *Sociobiology: the new synthesis*, 25th anniversary ed. Published by The Belknap Press of Harvard University Press. p334-335.

WHAT DO WOMEN THINK?

Fallon, A. E. & Rozin, P. (1985). Sex differences in perceptions of desirable body shape. *Journal of Abnormal Psychology* 94(1);102-105.

Jacobi, L. & Cash, T. F. (1994). In pursuit of the perfect appearance: discrepancies among self-ideal percepts of multiple physical attributes. *Journal of Applied Social Psychology* 24(5);379.

Klein, A. (1990). Little big men: hustling, gender narcissism and bodybuilding subculture. In Messner, M. & Sabo, D. (eds). *Sport, Men and the Gender Order: Criticial Feminist Perspectives*. Published by Human Kinetics, pp127-139.

Lynch, S. M. & Zellner, D. A. (1999). Figure preferences in two generations of men: the use of figure drawings illustrating differences in muscle mass. *Sex Roles: A Journal of Research* 40(9-10);833-843.

Tantleff-Dunn, S. & Thompson, J. K. (1995). Romantic partners and body image disturbance: further evidence for the role of perceived-actual disparities. *Sex Roles: A Journal of Research* 33(9-10);589-605.

Pope, H. G., Gruber, A. J., Mangweth, B., Bureau, B., Decol, C., Jouvent, R. & Hudson, J. I. (2000). Body image perception among men in three countries. *The American Journal of Psychiatry* 157(8);1297-1301.

Pope, H. G., Phillips, K. A. & Olivardia, R. (2002). *The Adonis Complex: how to identify, treat and prevent body obsession in men and boys.* Published by Simon & Schuster Inc.

SOCIAL AND PHYSICAL DANGERS OF BODYBUILDING

American Psychiatric Association. (2000). *Diagnostic and Statistical Manual of Mental Disorders*, 4th ed.

Berman, J. A., Setty, A., Steiner, M. J., Kaufman, K. R. & Skotzko, C. (2006). Complicated hypertension related to the abuse of ephedrine and caffeine alkaloids. *Journal of Addictive Diseases* 25(3);45-48.

Boothby, B. E. (2004). *What's the big deal about protein?* Harvard University Health Sciences Nutrition Services.

Cauter, E., Plat, L., Scharf, M. B., Leprolt, R., Cespedes, S., L'Hermite-Baleriaux, M. & Copinschi, G. (1997). Simultaneous stimulation of slow-wave sleep and growth hormone secretion by gamma-hydroxybutyrate in normal young men. *Journal of Clinical Investigation* 100(3);745-753.

Funk, R. (2006). Productive orientation and mental health. *Fromm Forum* 10;7-11.

Goertzen, M., Schoppe, K., Lange, G. & Schulitz, K. P. (1989). Injuries and damage caused by excess stress in body building and power lifting. *Sportvertletz Sportschaden* 3(1);32-36.

Graham, A. S. & Hatton, R. C. (1999). Creatine: a review of efficacy and safety. *Journal of the American Pharmaceutical Association* 39(6);803-810.

International Narcotics Control Board. (2006). *List of precursors and chemicals frequently used in the illicit manufacture of narcotic drugs and psychotropic substances under international control.* Annex to Form D, 10th ed. UN Convention against Illicit Traffic in Narcotic Drugs and Psychotropic Substances (1988).

Juhn, M. S. & Tarnopolsky, M. (1998). Potential side effects of oral creatine supplementation: a critical review. *Clinical Journal of Sport Medicine* 8(4);298-304.

Yoshizumi, W. M. & Tsourounis, C. (2004). Effects of creatine supplementation on renal function. *Journal of Herb and Pharmacotherapy* 4(1);1-7.

ABUSE OF ANABOLIC-ANDROGENIC STEROIDS

Billote-Domingo, K., Mendoza, M. T. & Torres, T. T. (1999). Catheter-related urinary tract infections: incidence, risk factors and microbiologic profile. *Philippine Society of Microbiology and Infectious Diseases* 28(4);133-138.

Burniston, J. G., Ng, Y., Clark, W. A., Colyer, J., Tan, L. & Goldspink, D. F. (2002). Myotoxic effects of clenbuterol in the rat heart and soleus muscle. *Journal of Applied Physiology* 93;1824-1832.

DuRant, R. H., Rickert, V. I., Ashworth, C. S., Newman, C. & Slavens, G. (1993). Use of multiple drugs among adolescents who use anabolic steroids. *New England Journal of Medicine* 328;922-926.

Embleton, P. & Thorne, G. (1998). *Anabolic Primer: An information-packed reference guide to ergogenic aids for hardcore bodybuilders.* Published by Robert Kennedy Publishing, p52

Fudala, P. J., Weinreib, R. M., Calarco, J. S., Kampman, K. M. & Boardman, C. (2003). An evalution of anabolic-androgenic steroid abusers over a period of 1 year: seven case studies. *Annals of Clinical Psychiatry* 15(2);121-130.

Grunfeld, C., Kotler, D. P. Dobs, A., Glesby, M., Bhaslin, S. (2006). Oxandrolone in the treatment of HIV-associated weight loss in men: a randomized, double-blind, placebo-controlled study. *Journal of Acquired Immune Deficiency Syndrome* 41(3),304-314.

Kanayama, G., Gruber, A. J., Pope, H. G., Borowiecki, J. J. & Hudon, J. I. (2001). Over-the-counter-drug use in gymnasiums: an underrecognized substance abuse problem? *Psychotherapy and Psychosomatics* 70(3);137-140.

Kearns, C. F. & McKeever, K. H. (2002). Clenbuterol diminishes aerobic performance in horses. *Medicine and Science in Sports and Exercise* 34(12);1976-1985.

Korkia, P. (1996). Use of anabolic steroids has been reported by 9% of men attending gymnasiums. *British Medical Journal* 212;1009.

Mayer, M. & Romain, O. (2001). Sciatic paralysis after a buttock intramuscular injection in children: an ongoing risk factor. *Archives of Pediatrics* 8(3);321-323.

Mouridsen, H. T. (2006). Incidence and management of side effects associated with aromatase inhibitors in the adjuvant treatment of breast cancer in postmenopausal women. *Current Medical Research and Opinion* 22(8);1609-1621.

Parkinson, A. B. & Evans, N. A. (2006). Anabolic androgenic steroids: a survey of 500 users. Medicine and Science in Sports and Exercise 38(4);644-651.

Pope, H. G. & Katz, D. L. (1994). Psychiatric and medical effects of anabolic-androgenic steroid use. A controlled study of 160 athletes. *Archives of General Psychiatry* 51(5);375-382.

Rannazzisi, J. T. (2004). *Statement before the House Committee on the Judiciary Subcommittee on Crime, Terrorism and Homeland Security.* DEA Congressional Testimony; Anabolic Steroid Control Act of 2004.

Roberts, A. (2004). *Beyond Steroids.* E-book published by http://www.elitefitness.com/

Shroeder, E. T., Vallejo, A. F., Zheng, L., Stewart, Y., Flores, C., Nakao, S., Martinez, C. & Sattler, F. R. (2005). Six-week improvements in muscle mass and strength during androgen therapy in older men. *The Journals of Gerontology. Series A, Biological Sciences and Medical Sciences* 60(12),1586-1592.

Stilger, V. G. & Yesalis, C. E. (1999). Anabolic-androgenic steroid use among high school football players. *Journal of Community Health* 24(2);131-145.

Tricker, R., Casaburi, R., Storer, T. W., Clevenger, B., Berman, N., Shirazi, A. & Bhasin, S. (1996). The effects of supraphysiological doses of testosterone on angry behavior in healthy eugonadal men – a clinical research center study. *Journal of Clinical Endocrinology & Metabolism* 81;3754-3758.

Wynaden, D., Landsborough, I., Chapman, R., McGowan, S., Lapsley, J. & Finn, M. (2005). Establishing best practice guidelines of intramuscular injections in the adult: a systematic review of the literature. *Contemporary Nurse* 20(2);267-277.

PSYCHIATRIC BODY IMAGE DISORDERS

American Psychiatric Association. (2000). *Diagnostic and Statistical Manual of Mental Disorders*, 4th ed.

Barry, A. & Lippman, S. B. (1990). Anorexia nervosa in males. *Postgraduate Medicine* 87(8);161-165.

Chung, B. (2001). Muscle dysmorphia: a critical review of the proposed criteria. *Perspectives in Biology and Medicine* 44(4);565-574.

Cotterill, J. A. & Cunliffe, W. J. (1997). Suicide in dermatological patients. *British Journal of Dermatology* 137(2);246-250.

Crerand, C. E., Phillips, K. A., Menard, W. & Fay, C. (2005). Nonpsychiatric medical treatment of body dysmorphic disorder. *Psychosomatics* 46;549-555.

Goldfield, G. S., Blouin, A. G. & Woodside, D. B. (2006). Body image, binge eating and bulimia nervosa in male bodybuilders. *Canadian Journal of Psychiatry* 51(3);160-168.

Hitzeroth, V., Wessels, C., Zungu-Dirwayi, N., Oosthuizen, P. & Stein, D. J. (2001). Muscle dysmorphia: a South African sample. *Psychiatry and Clinical Neurosciences* 55(5);521-523.

Mangweth, B., Hausmann, A., Walch, T., Hotter, A., Rupp, C. I., Biebl, W., Hudson, J. I. & Pope, H. G. (2003). Body fat perception in eating-disordered men. *International Journal of Eating Disorders* 35(1);102-108.

Olivardia, R., Pope, H. G. & Hudson, J. I. (2000). Muscle dysmorphia in male weightlifters: a case-control study. *American Journal of Psychiatry* 157(8);1291-1296.

Olivardia, R., Pope, H. G., Borowiecki, J. R. & Cohane, G. H. (2004). Biceps and body image: the relationship between muscularity and self-esteem, depression and eating disorder symptoms. *Psychology of Men & Masculinity* 5(2);112-120.

Penzel, F. (2000). *Obsessive compulsive disorders: a complete guide to getting well and staying well.* Published by Oxford University Press.

Perugi, G., Akiskal, H. S., Giannotti, D., Frare, F., Di Vaio, S. & Cassano, G. B. (1997). Gender-related differences in body dysmorphic disorder (dysmorphophobia). *Journal of Nervous and Mental Disease* 185(9);578-582.

Phillips, K. A., McElroy, S. L., Keck, P. E., Pope, H. G. & Hudson, J. I. (1993). Body dysmorphic disorder: 30 cases of imagined ugliness. *American Journal of Psychiatry* 150;302-308.

Phillips, K. A. & Diaz, S. F. (1997). Gender differences in body dysmorphic disorder. *Journal of Nervous and Mental Disease* 185(9);570-577.

Phillips, K. A. (2000). Pharmacologic treatment of body dysmorphic disorder: a review of empirical data and a proposed treatment algorithm. *Psychiatric Clinics of North America* 7;59-82.

Phillips, K. A. & Castle, D. J. (2001). Body dysmorphic disorder in men: psychiatric treatments are usually effective. *British Medical Journal* 323(7320);1015-1016.

Pope, H. G., Katz, D. L. & Hudson, J. I. (1993). Anorexia nervosa and "reverse anorexia" among 108 male bodybuilders. *Comparative Psychiatry* 34(6);406-409.

Pope, H. G., Gruber, A. J., Choi, P., Olivardia, R. & Phillips, K. A. (1997). An underrecognized form of body dysmorphic disorder. *Psychosomatics* 38;348-347.

Pope, H. G., Phillips, K. A. & Olivardia, R. (2002). *The Adonis Complex: how to identify, treat and prevent body obsession in men and boys.* Published by Simon & Schuster Inc.

Pope, C. G., Pope, H. G., Menard, W., Fay, C., Olivardia, R. & Phillips, K. A. (2005). Clinical features of muscle dysmorphia among males with body dysmorphic disorder. *Body Image* 2(4);395-400.

Ziora, K., Oswiecimska, J., Szalecki, M., Geisler, G., Broll-Waska, K., Kwiecien, J., Gorczyca, P., Lukasik, M., Franiczek, W., Stojewska, M. & Dyduch, A. (2006). Anorexia nervosa in boys and men. *Wiadomości Lekarskie* 59(5-6);352-358.

CONCLUSION

Chapple, A., Ziebland, S. & McPherson, A. (2004). Qualitative study of men's perceptions of why treatment delays occur in the UK for those with testicular cancer. British Journal of General Practice 54(498);25-32.

Fairholt, F. W. (1860). *Costume in England: a history of dress to the end of the eighteenth century.* Published by Chapman & Hall, p216-217.

Pope, H. G., Gruber, A. J., Choi, P., Olivardia, R. & Phillips, K. A. (1997). An underrecognized form of body dysmorphic disorder. *Psychosomatics* 38;348-347.

Pope, H. G., Phillips, K. A. & Olivardia, R. (2002). *The Adonis Complex: how to identify, treat and prevent body obsession in men and boys.* Published by Simon & Schuster Inc.

Appendices

Appendix I

Results of an anonymous online questionnaire marketed primarily at men exhibiting a high degree of muscularity on Venice Beach or at Muscle Beach, Los Angeles, California. N = 228.

Question	Total	Answer		
How many hours a week do you spend lifting weights or engaging in any other activity with the primary intention of gaining muscle?	225	Various	225	Avg 7.2 hours
How many hours a week do you spend on associated activities such as shopping, preparing special diets, travelling to the gym etc?	226	Various	226	Avg 7.0 hours
How many hours a week do you spend on other appearance related activities (such as tanning) over and above what might be considered normal?	224	Various	100	45%
		None	124	55%
How much money do you spend improving your appearance each week? Please include gym membership, supplements, special diets etc.	223	Various	223	Avg $56

REASONS FOR WEIGHT LIFTING

Please assign a percentage to each of the following reasons to indicate its importance in your decision to lift weights. Please also note that percentages assigned should add up to 100%	217	Attracting potential partners	7439	34.3%
		Achieving status	6320	29.2%
		among peers	7941	36.6%
		Improving health or ability for a particular sport		

COMMITMENT TO WEIGHT LIFTING

Weight lifting sometimes prevents me from	217	True	116	53.5%

socializing with friends (i.e. because of a work out schedule, avoiding alcohol etc)		False	101	46.5%
I would be willing to give up three years of my life to possess – or maintain – my ideal body	217	True	137	63.1%
		False	80	36.9%

USE OF BODYBUILDING SUPPLEMENTS

Please indicate the following bodybuilding supplements which you have used for a period equal to or exceeding two months	228	Protein bars or shakes	183	80.3%
		Creatine monophosphate	102	44.7%
		Artificial sources of amino acids (i.e. purified glutamine)	81	35.5%
		Natural testosterone boosters	23	10.1%
		Natural growth hormone boosters	10	4.4%
		Anti-oestrogen supplement	8	3.5%
		Pro-hormones	9	3.9%
		Ephedrine based fat loss aids	7	3.1%
		Other	41	18%
		None	30	13.2%

Appendix II

Results of an anonymous online questionnaire marketed to users of anabolic-androgenic steroids (AAS) at various online support and advice forums. N = 91.

Question	Total	Answer		
PERSONAL DETAILS				
Year of birth	90	Various		
Nationality	89	American	67	75.3%
		British	5	5.6%
		Australian	0	0.0%
		Canadian	13	14.6%
		Other	4	4.5%
Gender	89	Male	89	100%
		Female	0	0%
Please indicate whether or not you are - or are hoping to be - a competition bodybuilder	90	I am a competition bodybuilder	8	8.9%
		I have aspirations to become a competition bodybuilder	23	25.6%
		I do not intend to become a competition bodybuilder	59	65.6%
REASONS FOR USING ANABOLIC STEROIDS				
Age at first use of an anabolic steroid	89	Various		Avg 19
How were you first introduced to anabolic steroids?	89	I was offered anabolic steroids in a gym environment	2	2.2%
		I was offered	8	9.0%

		anabolic steroids outside of a gym environment		
		I actively sought out anabolic steroids	58	65.2%
		I was introduced to anabolic steroids by a friend	21	2.4%
What is your USUAL source of anabolic steroids?	89	An internet source	50	56.2%
		A gym contact	11	12.4%
		A friend	28	31.5%
Please assign a percentage to each of the following reasons to indicate its importance in your decision to use anabolic steroids. Please also note that percentages assigned should add up to 100%	87	Achieving status among peers	2255	27.8%
		Improving health or ability for a particular sport	3949	48.8%
		Attracting potential partners	2496	30.8%

USE OF ANABOLIC STEROIDS

Please indicate the ROUTE OF ADMINISTRATION used for your first steroid cycle	91	Oral	24	26.4%
		Sub-lingual	0	0.0%
		Intramuscular injection	58	63.7%
		Mixed oral and I/M	9	9.9%
Please indicate the ROUTE OF ADMINISTRATION used for your most recent steroid cycle	91	Oral	13	14.3%
		Sub-lingual	0	0.0%
		Intramuscular	65	71.4%

		ar injection		
		Mixed oral and I/M	13	14.3%
Please indicate which of the following steroids and associated compounds that you have completed at least one full cycle of	91	Aldactone	1	1.1%
		Anabolicum Vister	0	0.0%
		Anadrol	14	15.4%
		Andriol	2	2.2%
		Anavar	19	20.9%
		Arimidex	13	14.3%
		Cheque Drops	0	0.0%
		Clomid / Nolvadex	45	49.5%
		Cytomel	14	15.4%
		Danatrol	0	0.0%
		Deca-Durabolin	38	41.8%
		Dianabol	52	57.1%
		Equipoise	29	31.9%
		Esiclene	0	0.0%
		Finaplix	19	20.9%
		Halotestin	5	5.5%
		Lasix	0	0.0%
		Laurabolin	1	1.1%
		Masteron	9	9.9%
		Megagrisevit Mono	0	0.0%
		MENT	0	0.0%
		Methandriol	14	15.4%
		Methyltestosterone	0	0.0%
		Metribolone	0	0.0%
		Miotolan	0	0.0%
		Nilevar	0	0.0%
		Orabolin	0	0.0%
		Parabolan	2	2.2%
		Primobolan	4	4.4%
		Proviron	11	12.1%
		Stenbolone	0	0.0%
		Steranabol	1	1.1%
		Sustanon 250	25	27.5%
		Teslac	0	0.0%
		Testosterone Cypionate	24	26.4%
		Testosterone	59	64.8%

		Enanthate		
		Testosterone Propionate	27	29.7%
		Testosterone Suspension	5	5.5%
		Winstrol / Stromba	22	24.2%

CONSEQUENCES OF ANABOLIC STEROID USE

Please list any side-effects which you have experienced and that you attribute to anabolic steroid use	80	Temper tantrums or unusual aggressiveness	22	27.5%
		Confusion	0	0.0%
		Sleeping disorders	20	25%
		Steroid dependence	3	3.8%
		Paranoia	2	2.5%
		Hallucinations	0	0%
		Depression	9	11.2%
		Gynechomastia (growth of "female" breast tissue)	10	12.5%
		Testiciular atrophy (reduced testicular size)	63	78.8%
		Acne	55	68.8%
		Hair loss	10	12.5%
		None	11	13.8%
Please list any other pharmaceutical products which you have taken immediately prior to, during or after a steroid cycle	89	Human chorionic gonadotrophin (hCG)	22	24.7%
		Anti-estrogens	68	76.4%
		Anti-depressants	7	7.9%
		Human Growth Hormone	9	10.1%

		(hGH)		
		Insulin	6	6.7%
		Erythropoietin	1	1.1%
		Clenbuterol	33	37.1%
		None	7	7.9%
Please list any other illegal drugs which you began taking BEFORE your first steroid cycle	89	Amyl nitrate (poppers)	2	2.6%
		Cocaine	8	8.9%
		Crack cocaine	2	2.6%
		Crystal Meth	2	2.6%
		Ecstacy ('E')	10	10.7%
		Heroin	1	1.0%
		LSD (acid)	6	7.1%
		Marijuana (weed)	27	30.4%
		Speed	6	6.3%
Please list any other illegal drugs which you began taking AFTER your first steroid cycle	89	Amyl nitrate (poppers)	1	1.0%
		Cocaine	3	3.6%
		Crack cocaine	1	1.0%
		Crystal Meth	1	1.0%
		Ecstacy ('E')	2	2.7%
		Heroin	2	1.8%
		LSD (acid)	2	1.8%
		Marijuana (weed)	9	9.8%
		Speed	1	1.0%
Please indicate whether the following statement is true: "My source of anabolic steroids also sells illegal recreational drugs which have nothing to do with gaining muscle"	85	True	6	7.1%
		False	79	92.9%
Please list which of the following drugs (if any) that you have bought from a regular supplier of anabolic steroids	88	Amyl nitrate (poppers)	1	1.1%
		Cocaine	2	2.2%
		Crack cocaine	1	1.1%
		Crystal Meth	1	1.1%

	Ecstacy (E)	2	2.2%
	Heroin	2	2.2%
	LSD (acid)	1	1.1%
	Marijuana (weed)	1	1.1%
	Speed	1	1.1%
	None	85	96.6%